Marketing Practices in Educational Tourism

Practitioner-Led Approach to Strategy, Branding, and Innovation

Marketing Practices in Educational Tourism

Practitioner-Led Approach to Strategy, Branding, and Innovation

Zhiqi Tian

World Research Travel Organization (WRTO) Asia-Pacific Region, China

World Scientific

NEW JERSEY • LONDON • SINGAPORE • BEIJING • SHANGHAI • HONG KONG • TAIPEI • CHENNAI • TOKYO

Published by

World Scientific Publishing Co. Pte. Ltd.
5 Toh Tuck Link, Singapore 596224
USA office: 27 Warren Street, Suite 401-402, Hackensack, NJ 07601
UK office: 57 Shelton Street, Covent Garden, London WC2H 9HE

Library of Congress Control Number: 2025045109

British Library Cataloguing-in-Publication Data
A catalogue record for this book is available from the British Library.

MARKETING PRACTICES IN EDUCATIONAL TOURISM
Practitioner-Led Approach to Strategy, Branding, and Innovation

ISBN 978-981-98-2465-6 (hardcover)
ISBN 978-981-98-2466-3 (ebook for institutions)
ISBN 978-981-98-2467-0 (ebook for individuals)

For any available supplementary material, please visit
https://www.worldscientific.com/worldscibooks/10.1142/14628#t=suppl

Desk Editors: Soundararajan Raghuraman/Yipeng Hu

Typeset by Stallion Press
Email: enquiries@stallionpress.com

About the Author

Tian Zhiqi is a faculty member at Guilin Tourism University in Guangxi, China. He holds a Ph.D. in Tourism Management from Sichuan University and a Master's degree from The Hong Kong Polytechnic University. He currently serves as the Chief Operating Officer for the Asia-Pacific region of the World Research Travel Organization (WRTO) and is also a marketing expert at the Think Tank Center of the China Tourism Hotline Alliance.

With a strong background in the tourism industry, Tian previously led key projects at Ctrip and Tongcheng Travel, gaining hands-on experience in destination branding and tourism product development. His main research interests include digital marketing in cultural and tourism sectors, destination branding, and the planning of research-based educational travel.

Tian has spearheaded several major study travel initiatives, such as:

- Integrated Study Travel Planning for Leshan, Sichuan
- Red Tourism Product Development for Chishui
- Planning Services for the Malaysia Study Travel Summit
- International Study Travel Conference Planning for Singapore

He has also contributed to the development of both domestic and international guidelines for study travel programs.

Forward by Ma Yong

Read thousands of books, travel thousands of miles. Since ancient times, educational tourisms have been essential for Chinese students to broaden their horizons and deepen their knowledge. Since the issuance of the *Opinions on Promoting Educational Tourisms for Primary and Secondary School Students* by the Ministry of Education and ten other departments in 2016, educational tourisms in China have proliferated and shown strong momentum. In 2019, the Ministry of Education added the "Educational tourism Management and Services" major to the list of diploma-level programs in vocational colleges. In 2021, the Ministry of Culture and Tourism released the *14th Five-Year Plan for Cultural Industry Development*, which proposed developing new educational tourism products that combine cultural experiences, technological innovation, knowledge dissemination, leisure, and parent-child interaction.

With the rapid growth of this new industry, there is an urgent need for many skilled professionals. Therefore, it has become necessary to develop and publish a high-level, high-quality textbook series tailored to meet the evolving needs of the industry.

Huazhong University of Science and Technology Press, directly under the Ministry of Education and part of a national "Double First-Class" university, partnered with the World Research Tourism Organisation to launch a key project titled *Curriculum System Construction and Textbook Development Based on Talent Training Goals in Educational Tourism Management*. The aim was to develop a textbook series that combines international perspectives with Chinese characteristics, merges academic theory with practical guidance, and serves higher education institutions and industry professionals. The project was officially announced at the 2020 World Educational tourism Conference, accompanied by an open call for contributions. It received enthusiastic responses from professors, scholars, and subject leaders at more than 40 well-known universities and experts at nearly a hundred educational tourism bases, service agencies, and leading primary and secondary school teachers across China.

The results of this initiative have culminated in this series of textbooks. The initial plan includes nine titles: *Introduction to Educational Tourisms*, *Overview of Educational Tourism Resources*, *Educational Tourism Curriculum Development and Management*, *Practical Skills for Educational Tourism Tutors*, *Operations and Management of Educational Tourism Camps and Bases*, *Educational Tourism Product Design*, *Educational Tourism Project Development and Operation*, *Educational Tourism Marketing*, and *Educational Tourism Safety Management*. These books comprehensively cover all the significant aspects of today's educational tourism industry.

Several distinctive features mark this series. The contributing authors come from various regions across China. They possess global perspectives and rich international teaching experience and serve as leaders in universities or educational tourism institutions. They incorporate advanced international practices, unique Chinese values, national sentiment, and hands-on insights in writing. The editorial team is led by Professor Ma Yong, a national-level expert under the Central Organisation Department's High-level Talent Support Program and Vice Chairman of the Ministry of Education's Tourism Management Teaching Steering Committee, alongside Professor Yang Zhenzhi, Chairman of the World Research Tourism Organisation. Each book is edited by a team comprising deans, professors, and subject leaders from renowned universities such as Sichuan University, Hubei University, Fujian Normal University, and others, as well as heads of educational tourism camps and service providers. With their profound teaching and practical experience, and in alignment with guidance from the Ministry of Education and Culture and Tourism, they ensure the series' academic authority, accuracy, and forward-thinking approach.

Each volume in this series is the product of collaborative authorship between the academic community and industry. In addition to a sound theoretical foundation, the books include a wide range of practical and classic case studies. Real-world work processes guide their structure, and each book contains well-designed teaching projects and tasks, ensuring a strong link between theory and practice.

Huazhong University of Science and Technology Press has built a dedicated online resource platform to further support educators. It provides teaching plans, slides, question banks, case libraries, answer keys, instructional videos, and more. Vertically, it supports the complete teaching ecosystem, from course development and practice management to student feedback and class administration. These resources create a multidimensional, interactive learning experience integrating digital and print formats.

As a newly established discipline and a fast-developing industry, educational tourism management and services are entering a period of accelerated growth. Hopefully, this textbook series will nourish students and contribute meaningfully to developing China's educational tourism sector. We also sincerely invite more scholars to join us in this effort.

Ma Yong
May 2022

Forward by Yang Zhenzhi

This textbook series is the outcome of the key research project by the World Research Tourism Organisation titled "*Curriculum System Development and Textbook Compilation Based on the Talent Training Goals of the Research Tourism Major.*"

In China, research-based educational tourisms are flourishing. Government departments at all levels, parents, schools, students, and the general public are all eagerly anticipating the further development of this field. Whether it was the ancient Chinese tradition of educational tourism or the Western *Grand Tour*, centuries of experience have consistently proven that returning to the outdoors and learning in nature is one of the most effective ways to enhance an individual's overall competence.

In modern China, research tourism is still in its infancy. Drawing on more than a hundred years of nature-based education experience from developed Western countries, this series of textbooks aims to establish a research tourism education system with distinct Chinese characteristics. It also seeks to clarify misunderstandings in the field, whether related to the concept, the philosophy, or the practical operation of educational tourisms. This goal has shaped the starting point of our textbook development.

Thus, from the beginning, we set a clear guiding principle for this textbook series: an international perspective with Chinese characteristics, a strong emphasis on practice and operation, and integrating theory with application, achieving the unity of knowledge and action. In selecting the authors, we invited professionals who understand the national context of China and the global landscape of research tourism. We also encouraged them to assess international trends in nature education and draw on global case studies. Furthermore, we brought together academic researchers from universities and practitioners from the frontlines of research tourism enterprises to ensure a comprehensive approach. Integrating theory and practice is one of the series' defining features.

This series can serve as textbooks in colleges and universities, especially for vocational institutions offering the *Research Tourism Management and Services* major.

The World Research Tourism Organisation places great importance on the value of research tourism in personal growth and character development. It advocates for introducing research tourism from an early age, starting with kindergarten children and continuing through primary and secondary school. The ultimate goal is to improve an individual's overall competence and truly embody the principle of learning through doing. Research tourism encourages students to step out of the classroom and return to nature, to engage directly with the natural world. It emphasises real experience and practical learning in nature. It opposes the idea of merely moving from one classroom to another or continuing the traditional method of rote instruction in museums or outdoor settings. Research tours without hands-on experience or meaningful action cannot fulfil their intended purpose.

We hope this series of textbooks will contribute meaningfully to the growing field of research tourism in China and bring value to readers, especially parents. This is our humble offering to society.

Yang Zhenzhi
May 2022

Preface

Educational tourism is the integration of "education + tourism." Against the backdrop of China's "double reduction" policy introduced by the Ministry of Education, the education and tourism industries have gained new momentum and creativity. Educational tourism has thus entered a phase of significant development opportunities. However, there are still many challenges in its practical market operations, such as an unbalanced product structure, insufficient educational focus, unclear market segmentation, vague business models, limited marketing approaches, and unsafe operational standards—all of which hinder the healthy growth of the educational tourism market.

This book is led and compiled by experienced frontline industry practitioners, incorporating numerous practical case studies. It offers in-depth discussions on core topics, including market supply analysis of educational tourism products, positioning, brand building, channel marketing, new media marketing, and the application of intelligent marketing platforms in educational tourism. The deep integration of educational tourism with the internet is bound to drive digital transformation, and marketing in this field will gradually evolve into more refined, data-driven approaches. At the same time, this book emphasises the integration of forward-looking international perspectives with Chinese characteristics, as well as the fusion of practical educational tourism with educational principles. It is highly targeted at developing capabilities for educational tourism base operators, institutions, and students in product design, brand building, and marketing. We hope this book can serve as a valuable reference for all professionals in educational tourism.

Marketing educational tourism possesses its unique features, yet it remains in the developmental stage of the industry. To enhance student understanding, each chapter begins with a case study to guide students into exploratory learning and concludes with exercises to reinforce the integration of theory and practice. The content is structured around real-world examples to improve students' practical skills while deepening their theoretical knowledge.

The division of writing responsibilities for this book is as follows: Mr. Tian Zhiqi, Chief Operating Officer of the World Educational Tourism Organisation (Asia-Pacific), serves as the chief editor. He was responsible for designing the textbook outline, chapters, and structure and compiling the final manuscript. He authored Chapter 6 (Building Educational Tourism Brands and Brand Strategies) and Chapter 10 (Exploring the Application of Smart Marketing Platforms in Educational Tourism). He co-authored Chapter 11 (New Media Marketing for Educational Tourism) with Li Yuemei, a nationally recognised Chinese language teacher from Gulang County No. 1 Primary School in Gansu. Song Shuxia from the Dunhuang Academy wrote Chapter 7 (Market Segmentation and Target Market Strategies for Educational Tourism). Wang Xiaoyan from Hunan Jinhua International Travel Agency (IE International Education) wrote Chapter 4 (Educational Tourism Market Customer Analysis). Huang Fen, an expert from the World Educational Tourism Organisation's expert panel, authored Chapter 1 (Overview of Marketing). Tang Ningtao from the School of Tourism and Geography at Shaoguan University authored Chapters 2 and 3 (Overview of the Educational Tourism Market and Market Environment Analysis for Educational Tourism, respectively). Liu Baogang, Chairman of the Educational Tourism Committee of the Xiamen Youth Development Education Association, and Lin Ying from Chengdu Laiye Cultural Tourism Information Technology Co., Ltd. co-authored Chapter 5 (Marketing Channels and Expansion Strategies for Educational Tourism). From Hailiang Education Management Group, Hou Xiaogang, Jiang Peng, Chu Xinwei, Chu Jiabin, Gong Shengxian, and Yu Caorong co-authored Chapters 8 and 9 (Market Research for Educational Tourism and Formulating and Implementing Marketing Plans for Educational).

This book received tremendous support from various institutions, including the Leshan Municipal Bureau of Culture, Radio, Television and Tourism, the Sanxing Village Archaeological Base, the Department of Tourism, Culture, Radio, Television and Sports of Hainan Province, the Department of Culture and Tourism of Henan Province, and the Luoyang Municipal Bureau of Culture, Radio, Television and Tourism. We express our heartfelt thanks to all of them.

We have made every effort to reflect domestic and international trends in the development of educational tourism, adopting a collaborative model between educational institutions and enterprises. The book incorporates cutting-edge concepts from leading domestic educational tourism enterprises and is tailored to the needs and characteristics of vocational education students. Compiled under the guiding principles of combining theory with practice and applying what one has learned, this book is the result of our dedicated effort. However, given the ongoing evolution of the educational tourism market and its unique nature, there is still much to explore. Due to the authors' limited capabilities, the book's shortcomings are inevitable. We welcome any feedback from readers and will actively make improvements to better support the growth of educational tourism education.

Luoyang
August 2022

Contents

Chapter 1

Marketing Overview

Learning Objectives

1. Understand the concepts and elements of the market and marketing.
2. Understand the marketing environment and marketing philosophies.
3. Understand mainstream marketing theories.
4. Understand the concepts, elements, and environment of tourism marketing.
5. Master the basic principles of tourism market segmentation.
6. Master the educational tourism marketing concept and its industry chain structure.

Knowledge Structure

- **Overview of Marketing**
  - Understanding Marketing
    - Market concepts and elements
    - Marketing concepts and elements
    - Marketing products
    - Marketing philosophies
    - Marketing environment
    - Marketing theories
  - Understanding Tourism Marketing
    - Concept of tourism products
    - Concept of tourism marketing

  - Elements of tourism marketing
  - Characteristics of tourism marketing
  - Tourism marketing environment
  - Tourism market segmentation
- Understanding Educational Tourism Marketing
  - Concept of educational tourism marketing
  - Industry chain of educational tourism marketing

Learning Focus

1. The concepts, core marketing elements, environment, and key marketing theories.
2. The concepts and elements of tourism marketing and the basic principles of tourism market segmentation.
3. The concept of educational tourism marketing and its industry chain structure.

Contents

1. Case Introduction

IP Marketing: Igniting the Era of Traffic

IP marketing harnesses the powerful influence of prominent intellectual property (IP) brands to attract fans, thereby generating traffic and monetising value. The brand and its IP must establish a strong content-based connection and focus on integrated content marketing for effective outcomes. This strategy shifts from traditional commercial detachment to an entertainment-driven model, enhancing the fan experience and boosting commercial sales.

Taking the Palace Museum's cultural IP as an example, "Palace Taobao" is an artistic, creative IP developed by the Palace Museum to market its merchandise. Tencent skilfully employed internet thinking and leveraged the Palace Museum's compelling traffic-generating IP by launching a quirky and retro-themed campaign. They amalgamated time travel and rap elements into an H5 digital experience titled "Travel Through the Palace to See You," which quickly went viral. The seamless integration of H5 with short videos and the playful combination of rap and visuals created a surprising and highly engaging experience. The application of cutting-edge frontend technologies was spot on, rendering it an attention-grabbing interactive success. Within one day of its launch, the campaign exceeded 3 million views, achieving a phenomenal brand promotion result.

1.1. *Analysis & reflection*

(a) Please provide a brief interpretation of IP marketing.
(b) Share at least one tourism-related IP marketing case with your classmates and describe the most remarkable detail that caught your attention.

2. Understanding Marketing

2.1. *Market: concept and elements*

From the perspective of marketing science, a market is the aggregate of actual and potential buyer demand for a specific product or service. It refers to a group of consumers with particular needs and wants and the purchasing power to satisfy those wants. The market comprises three elements.

- Population — the size of the group with a common need.
- Purchasing Power — the financial ability to buy products.
- Desire to Purchase — the motivation to satisfy a need or want.

Formula:
Market = Population + Purchasing Power + Desire to Purchase

1. **Population**
   This primarily refers to the size of the population—the larger the population, the greater the market for a product. It also includes the number of individuals who share a common demand for a specific product: the target consumers that an enterprise can serve. The more numerous this group is, the larger the potential market becomes, and the better it supports the survival and development of the enterprise. Consequently, the population determines the size of the market.
2. **Purchasing Power**
   Purchasing power refers to an individual's ability to pay for the required goods or services. This capability is primarily determined by a person's income level, followed by the cost of goods (price levels) and, finally, their credit capacity.
3. **Desire to Purchase**
   The desire to purchase refers to the want and need for a specific product. This desire stems from both the physiological and psychological needs of the individual.

These three elements of the market—population, purchasing power, and desire to buy—are interconnected and mutually restrictive. Population is the fundamental element of market composition. The larger the population, the greater the actual and potential consumer demand. When the population is fixed, purchasing power becomes a key factor determining market size; the level of purchasing power directly influences the market's capacity. The desire to buy is the driving force behind consumer buying behaviour—it is the force that transforms potential purchasing power into actual purchasing actions, making it a critical condition for market activity.

2.2. *Marketing: concept and core elements*

The renowned American marketing scholar Philip Kotler defines marketing as:

> "A social and managerial process through which individuals and groups obtain what they need and want by creating, offering, and exchanging products and value with others."

Marketing is not only about meeting consumer needs, but also about determining what kind of products can meet those needs, and how they can be fulfilled, specifically, through what type of exchange, when and where the exchange occurs, and who facilitates the connection between the product and the consumer.

Therefore, the core elements of marketing should include:

(1) Needs, and the associated wants and demands
(2) Products, and their associated value, effectiveness, and satisfaction
(3) Exchange, and its associated transactions and relationships

2.2.1. *Needs, desires, and demand*

(a) Needs
Needs refer to the physiological and psychological requirements of consumers. For example, people need food, clothing, housing, security, belonging, respect, and self-actualisation to survive and thrive. These are fundamental human needs. Marketers cannot create such needs—they can only respond to them.

(b) Desires
Desires are the deeper, culturally shaped expressions of needs. Consumers from different social backgrounds have other desires. For instance, Chinese consumers may desire rice, French consumers bread, and American consumers hamburgers. Desires are influenced by social and institutional factors, as well as by occupation, groups,

and family. They evolve with changing social conditions. Marketers can influence consumers' desires by promoting specific products to shape preferences and guide choices.

(c) Demand

Demand refers to desires backed by purchasing power and willingness to buy. In other words, when a consumer has both the desire for a product and the ability to pay for it, that desire becomes actual demand.

For example, many people may desire luxury cars, but only those with the financial means to purchase them generate real demand. Therefore, marketers must understand how many people want their product and determine whether those people can pay for it.

2.2.2. *Products and the associated value, effectiveness, and satisfaction*

(a) Products

A product is anything offered to satisfy consumers' needs and desires. Products can be tangible or intangible, physical or non-physical.

- Tangible products serve as carriers for delivering services to consumers.
- Intangible products or services are delivered through other carriers such as people, places, activities, organisations, or ideas. Services can also be transmitted via tangible goods and other media.

For example, when we feel fatigued, we might enjoy music at a concert hall, relax in a park, participate in club activities, or embrace a new idea or philosophy.

(b) Effectiveness, Value, and Satisfaction

When consumers select the products they need, they primarily evaluate the effectiveness of each product in meeting their needs before making a decision. Effectiveness (utility) is the consumer's assessment of the total capability of a product to satisfy their needs.

How is the standard for a product's overall effectiveness (or an ideal product) determined?

For example, if a consumer needs to travel to a destination, the available options might include a car, train, high-speed rail, or aeroplane. These options form a product demand set. Consumers rank these products based on how well each meets their needs, from most to least satisfying, and choose the product that most closely matches their requirements.

When selecting products, consumers not only consider effectiveness (utility) but also take price into account. If consumers aim to maximise utility, they will not simply look at the absolute price of a product; instead, they will evaluate which product offers the most effective value for every unit of money spent.

2.2.3. *Exchange, transaction, and relationship (exchange, transaction, relationship)*

(a) Exchange

Even when people have needs and desires, and businesses have produced products, this alone does not constitute marketing. A product only becomes a marketing product through the process of exchange. Obtaining products through self-sufficiency, self-production, theft, or begging is not considered marketing. Only when two parties voluntarily exchange equivalent value, allowing both sides to acquire the products or services they need, does marketing truly occur. Thus, exchange is the core element of marketing.

To complete a successful exchange, the following five conditions must typically be met:

1. There must be at least two parties involved in the exchange.
2. Each party must have something that the other party desires.
3. Each party must be able to communicate with the other and deliver the goods or services.
4. Each party must be free to accept or reject the offer.
5. Each party must have the desire to engage in the exchange.

(b) Transaction

Exchange is a process, not a single event. If two parties are discussing and gradually reaching an agreement, they are said to be "in exchange." When an agreement is finalised, a transaction occurs.

A transaction refers to the actual value exchange between buyer and seller, usually involving money as the medium. (Whereas exchange can happen without money, such as in barter.)

A typical transaction includes:

- Two items of value (e.g., goods or services)
- Agreed-upon conditions
- A specific time and place
- Legal systems to uphold and enforce commitments if necessary

(c) Relationship
Relationship marketing extends beyond single transactions—it aims to build long-term, mutually beneficial connections between businesses and consumers, distributors, and other partners.

- Successful relationships are based on continuous commitment, high-quality products, excellent service, and fair pricing.
- Ongoing economic, technical, and social bonds strengthen them.

Many outstanding companies embody this concept. For example, before launching a new Accord model in the U.S., Honda engineers travelled to Los Angeles to study local highways, measure dimensions, and analyse driving conditions. They even built a replica U.S. highway in Japan to ensure the car matched American consumer needs. This deep understanding contributed significantly to the Accord's success in the American market.

Similarly, every staff member at Disneyland must embody the "Disney spirit," ensuring all visitors experience joy and wonder. Employees are trained never to respond "I don't know" to customer inquiries. This consistent, high-quality service builds long-term emotional connections with visitors, making them eager to return again and again.

2.3. *Marketing products*

Marketing products can be tangible goods or intangible services. According to the theory outlined in *Marketing Management*, products in marketing can generally be classified into the following ten categories:

(a) Goods
- Physical, tangible items such as food, automobiles, and refrigerators.

(b) Services
- Activities or benefits such as haircuts from barbers, beauty treatments from beauticians, or a restaurant that offers both food (goods) and service.

(c) Events
- Large-scale events like trade exhibitions, art performances, corporate celebrations, or global events like the Olympic Games.

(d) Experiences
- Combinations of products and services that provide consumers with extraordinary experiences, such as the immersive environment created by Disneyland.

(e) People
- Celebrity marketing is a typical example. Many stars, artists, and consultants successfully build strong personal brands and effectively market themselves.

(f) Places
- Locations marketed for tourism, living, or business, such as hotels, cities, or resort destinations.

(g) Properties
- This includes real estate and financial assets like stocks and securities.

(h) Organisations
- Various organisations, including universities, museums, art troupes, and non-profit groups, work to build and promote their brand image through marketing.

(i) Information
- The production, packaging, and distribution of information is a massive industry. Many intermediary and educational institutions depend heavily on the marketing of information.

(j) Ideas
- Every market offering carries an underlying idea or concept. Products and services often serve as carriers for promoting broader ideas or values, such as environmental protection or social innovation.

2.4. *Marketing philosophies*

Marketing philosophies are the fundamental guiding principles that direct a company's marketing activities. At its core, the marketing philosophy addresses how a company balances the interests of the enterprise, consumers, and society.

The evolution of marketing philosophy can be divided into two stages:

- Traditional marketing philosophies
- Modern marketing philosophies

2.4.1. *Traditional marketing philosophies*

(a) Production Concept

This concept emerged before the 1920s and is best summarised: "I sell whatever I produce." The production concept assumes that consumers prefer readily available and affordable products. Therefore, businesses should focus on improving production efficiency and distribution effectiveness, expanding production scale, and reducing costs to grow the market. As such, this philosophy emphasises production while neglecting market influence.

(b) Product Concept

This concept coexisted with the production concept and is best captured by the idea: "As long as the product is of good quality, it will sell." The product concept holds that consumers prefer high-quality, multi-functional, and distinctive products, so businesses should focus on improving product quality and continuously developing new products.

(c) Selling Concept

The selling concept emerged during the transition from a seller's market to a buyer's market in capitalist countries. Its core belief is: "Whatever I promote, you will buy." According to this concept, consumers typically do not actively seek out or purchase products, but are instead influenced by promotional efforts that stimulate demand. Therefore, consumers will buy more if a business works hard to sell a product.

2.4.2. *Modern marketing philosophy*

The modern marketing concept is a consumer-oriented philosophy of marketing management. It focuses on satisfying consumer needs and desires through integrated marketing strategies to achieve long-term business success. Its core beliefs can be summarised as:

- **"We produce what consumers need."**
- **"We sell what the market demands."**
- **"Where there is consumer demand, there is a marketing opportunity."**

Thus, modern marketing shifts the mindset from "Make a product and try to sell it" to "Identify a need and find a way to meet it." It is no longer about promoting what has already been produced but about creating products that can be sold because they meet genuine consumer demand.

The theoretical foundation of this concept is the idea of "consumer sovereignty," which asserts that the decision of what to produce lies not with the manufacturer or the government, but with the consumer. The consumer is the dominant force in the relationship between producers and consumers. Producers should organise their production based on consumer preferences and needs. Businesses can only achieve customer satisfaction and profitability by creating products that meet these needs. This perspective emerged and gained popularity under buyer's market conditions, where product supply exceeds demand.

2.5. *Marketing environment*

The marketing environment refers to all external forces and factors that impact a company's marketing activities. These external factors can create opportunities or threaten a company's marketing efforts. Depending on their relationship with the business, the marketing environment can be divided into two types: (a) Macro-environment and (b) Micro-environment.

2.5.1. *Macro-environment*

The **macro-environment** comprises major societal forces that **indirectly affect** a company's marketing activities. These include:

1. **Demographic Environment**
   Population size and structure significantly influence market scale and potential. Geographic distribution, age structure, gender ratio, ethnic composition, and family dynamics shape market trends.
2. **Economic Environment**
   The **economic environment** is another key factor influencing a company's marketing activities. It includes elements such as **income levels, consumer spending, savings and credit**, and the **overall level of economic development**.

   Among these, **consumer income** is significant. It refers to consumers' total monetary income through various channels, including **wages, bonuses, dividends, interest, commissions, and inheritances**. Changes in consumer income are primarily influenced by **national income** and serve as a significant factor in shaping **societal purchasing power**.

3. **Natural Environment**
   The natural environment comprises the physical factors that influence a company's development. When evaluating this environment, businesses should consider whether natural resources are becoming scarce, whether environmental pollution is worsening, whether government regulations and interventions concerning environmental protection are increasing, and whether the public's ecological awareness and demand for sustainability are continuing to rise.
4. **Technological Environment**
   Technology is a force of "creative destruction." This concept aptly captures the profound impact of technological advancement on business marketing. For example, new technological revolutions have shortened product life cycles, altered consumer shopping habits, and transformed how companies operate and manage their businesses.
5. **Political and Legal Environment**
   Under any social system, a company's marketing activities must be regulated and constrained by the political and legal environment. This environment constantly affects businesses, as all corporate activities operate within the boundaries set by political and legal frameworks.
6. **Socio-cultural Environment**
   The **socio-cultural environment** refers to conditions shaped by a society's existing **national characteristics, value systems, religious beliefs, lifestyles, customs and traditions, ethics and morality, educational levels, social groups,** and **social structures**.

2.5.2. *The microenvironment of marketing*

The microenvironment of marketing refers to the various forces that directly influence a company's ability to serve its customers. These include the company, suppliers, competitors, marketing intermediaries, consumers, and the public.

1. **The Company Itself**
   All internal organisational structures form a company's internal microenvironment at two levels. The first level is top management. The marketing department must operate within the authority defined by senior management, and its plans must receive top-level approval before implementation. The second level includes the company's other

functional departments. Since marketing operations are closely linked to the activities of other departments, effective marketing requires coordination and collaboration across the organisation to achieve desired results.

2. **Suppliers**

   Suppliers are businesses or individuals that provide the resources required for a company's production and operations. These resources include raw materials, components, equipment, energy, labour, intellectual services, and capital. These providers' stability, timeliness, pricing, and supply quality can significantly influence a company's marketing performance.

3. **Competitors**

   In any market that is not a monopoly, competitors are inevitable. Even when only one business visibly offers a product or service in a market, it is often difficult to conclude that no potential competitors exist. The presence of existing and potential competitors makes competition a constant force that companies must respond to in their marketing strategies.

4. **Marketing Intermediaries**

   Marketing intermediaries are businesses or individuals that help companies promote, sell, and distribute products to end customers. These include mediators, physical distribution organisations, marketing service agencies, and financial institutions. They play a vital role in connecting companies to their target markets and ensuring the smooth flow of goods and services.

5. **Consumers**

   Consumers determine whether a company's products or services are accepted, and to what degree, which affects the effectiveness of its marketing activities. Understanding and meeting the needs of consumers is the central focus of marketing. In particular, attention should be paid to the influence and expectations of different consumer segments.

6. **The Public**

   The public refers to groups and individuals who have a real or potential impact on a company's ability to achieve its marketing objectives. This includes external publics such as the media, government agencies, community groups, financial institutions, and internal publics like employees, shareholders, and management. How a company manages its relationships with these stakeholders is essential to its overall marketing environment.

2.6. *Marketing theories*

(1) 4Ps Marketing Theory

In 1960, Professor Jerome McCarthy of Michigan State University summarised the key marketing elements in his book *Basic Marketing* as four categories: Product, Price, Place, and Promotion.

In 1967, Philip Kotler established the 4Ps as the core of the marketing mix in his best-selling book *Marketing Management: Analysis, Planning, and Control*. Since all four terms begin with the letter "P" and are often used in conjunction with the strategy concept, this framework is commonly referred to as the "4Ps". The four components are as follows:

Product refers to developing a product's functionality, emphasising its unique selling proposition (USP). The primary focus is on what the product can do and how it meets the needs of consumers.

Price involves setting different pricing strategies based on the product's market positioning. Pricing decisions are grounded in the company's branding strategy, with particular emphasis on the perceived value and prestige of the brand.

Place refers to the channels through which a product reaches the consumer. Companies typically do not engage directly with consumers; instead, they focus on developing distributors and building sales networks. The connection between the company and the end consumer is facilitated through intermediaries, such as distributors and retailers.

Promotion is often narrowly interpreted as "sales promotion," but this view is overly simplistic. Promotion encompasses marketing activities, including brand advertising, public relations, and sales promotions. It represents the overall strategy for communicating value to consumers and influencing their purchasing decisions.

2.6.1. *4Cs Marketing theory*

The 4Cs Marketing Theory, proposed by American marketing expert Professor Robert F. Lauterborn in 1990, shifts the focus from the company's perspective to that of the consumer. It emphasises that businesses should prioritise customer satisfaction, followed by efforts to reduce the cost to consumers, enhance convenience in the buying process, and implement consumer-centred communication strategies rather than relying solely on traditional company-driven marketing channels. The 4Cs model includes the following elements:

1. **Customer's Need**
   The starting point is to understand, research, and analyse the needs of consumers, rather than beginning with what the company is capable of producing. Businesses must first gain insight into their target customers and offer products and services tailored to those needs. More importantly, the company must aim to deliver customer value beyond providing a product or service.
2. **Cost to the Consumer**
   This refers to what consumers are willing to pay to satisfy their needs and desires, not simply the price set by the company. Cost goes beyond the company's internal production costs or the traditional "price" in the 4Ps model. It also includes the consumer's total cost of purchase, such as time, effort, and psychological cost. Ideally, pricing should be below the consumer's psychological price threshold while still ensuring reasonable profitability for the company.
3. **Convenience**
   Products should be designed with ease of use and accessibility for the consumer in mind. Convenience refers to offering consumers maximum convenience in both purchasing and using the product. The 4Cs theory emphasizes that, when developing distribution strategies, companies should focus more on what is convenient for the consumer, rather than what is most convenient for the business itself.
4. **Communication**
   Consumer-centred communication is essential in modern marketing. By fostering interaction and dialogue, companies can continually integrate internal and external marketing efforts to align the consumer and business interests. In the 4Cs framework, communication replaces "promotion" in the 4Ps model, emphasising a two-way, relationship-building process over one-way promotional messaging.

2.6.2. *STP marketing theory*

Market segmentation was first introduced in 1956 by American marketing scholar Wendell Smith. It was later expanded and refined by Philip Kotler, who developed it into the now widely recognised STP model—an acronym for Segmentation, Targeting, and Positioning.

(1) Market Segmentation

Market segmentation refers to the process by which a company, through market research, divides the overall market for a product into smaller sub-markets, or segments, based on differences in consumer needs and desires, purchasing behaviour, and buying habits. Each resulting segment is known as a sub-market or market segment.

It is important to note that segmentation is not based on the product's characteristics, but rather on the differences among consumer groups. The foundation of market segmentation lies in consumer demand variability. Businesses must categorise the market by grouping consumers with similar desires and interests, forming distinct market segments. See Table 1.1 for the target market segmentation variables.

(i) Geographic Segmentation

Geographic segmentation refers to dividing the consumer market based on consumers' location and other geographic variables.

(ii) Demographic Segmentation

Demographic segmentation refers to dividing the consumer market according to demographic variables.

Table 1.1 For the target market segmentation variables.

Segmentation Variable	Specific Content
Geographic Variables	Country, region, city, rural area, land area, climate, terrain, transportation conditions, communication infrastructure, urban planning, etc.
Demographic Variables	Total population, population density, number of households, age, gender, occupation, ethnicity, education, religion, nationality, income, family, life cycle, etc.
Psychographic Variables	Personality, purchase motivation, value orientation, perceptions or preferences toward products and services, sensitivity to price changes, etc.
Behavioural Variables	Purchase timing, benefits sought, user status, brand loyalty, usage rate, readiness stage, attitudes, etc.

(iii) Psychographic Segmentation

Psychographic segmentation divides the consumer market based on lifestyle, personality, and other psychological variables.

(iv) Behavioural Segmentation

Behavioural segmentation refers to dividing the consumer market according to consumer behaviours, such as purchase timing, benefits sought, user status, usage frequency, brand (or store) loyalty, stage of readiness to purchase, and attitude toward the product.

(2) Target Market

The purpose of market segmentation is to provide a scientific basis for businesses to select their target markets. The choice of a target market determines who the company serves, what products it offers, the product quality or positioning, and how the products are sold. Whether this selection and strategy are accurate directly impacts the company's survival and development.

A target market refers to the segment or segments of the market that a company has decided to enter—markets identified through segmentation, which the company chooses to serve through its marketing activities.

There are three main strategies a company can adopt when selecting a target market:

I. Undifferentiated Marketing Strategy

An undifferentiated marketing strategy treats the entire market as a single target market, focusing only on the everyday needs of consumers while ignoring their differences. A company adopts one product, pricing model, and marketing approach to appeal to as many customers as possible. This strategy requires that the product have a distinctive internal and external style to gain broad consumer recognition and maintain relative stability. Its advantages include simplified production, quality control, mass production, and reduced production and sales costs. However, if other companies adopt the same strategy, it inevitably leads to intense competition.

II. Differentiated Marketing Strategy

A differentiated marketing strategy involves dividing the entire market into several segments and designing different products and tailored marketing strategies for each, to meet the distinct needs of each consumer group.

III. Concentrated Marketing Strategy

A concentrated marketing strategy means selecting one or a few segments from the overall market and focusing on specialised production and sales within those areas. This allows a business to leverage its strengths in a specific market to increase its market share. This strategy requires a deep understanding of the target market and is particularly suitable for small and **medium-sized enterprises**.

(3) Market Positioning

The concept of market positioning was introduced in the 1970s by American scholar Al Ries. Market positioning refers to the process by which a company, based on the competitive landscape and consumer preferences in the target market, creates a strong, distinctive identity for its product and communicates it effectively to consumers to gain their recognition.

The essence of market positioning is to differentiate a business from its competitors, so that consumers can clearly perceive the difference, thereby securing a unique position in the target audience's minds.

Traditionally, it was believed that positioning simply meant offering different products in each market segment, known as product differentiation. However, while related, market positioning and product differentiation differ. Positioning focuses on creating a unique identity for the product by highlighting specific features or attributes, such as performance, design, composition, packaging, shape, and quality, to form a distinctive market image.

Product differentiation is a means to achieve market positioning, not its entirety. Positioning emphasises differences in the product and aims to build a recognisable market image to win over consumers.

There are three primary market positioning strategies:

(1) Avoid-the-Strong Positioning Strategy

This strategy involves avoiding direct confrontation with major competitors by targeting market gaps or weak areas, introducing products not yet available in the target market, and developing new market segments.

(2) Head-to-Head Positioning Strategy

This is a direct competition strategy, where a company positions its product against the strongest competitor. Companies adopting this strategy must possess greater competitive strength than their rivals.

(3) Repositioning Strategy
If a competitor's product has a similar positioning and begins encroaching upon a company's market, or if consumer preferences shift toward a competitor, the company must consider repositioning its product. This involves changing consumer perceptions of the brand and establishing a new identity that resonates with the target audience.

3. Understanding Tourism Marketing

Tourism is a rapidly developing, promising, and large-scale emerging industry. It is also a highly interconnected and comprehensive sector. The development level of tourism has become an essential indicator of a country or region's economic progress, social advancement, and quality of life. The sustained, rapid, and healthy growth of the tourism industry plays an increasingly vital role in driving related industries, promoting openness, optimising the economic structure, creating employment opportunities, alleviating poverty, enriching people's lives, developing local economies, cultivating new sources of revenue, accelerating urbanisation, and enhancing social and cultural development.

3.1. *Concept of tourism product*

From the supply-side perspective, a tourism product refers to the total collection of goods and services provided by a tourism enterprise through the market that can satisfy the needs of tourists during a single tourism activity.

From the consumer-side perspective, a tourism product is a combination of tangible goods and intangible services that tourists can experience during a single tourism activity, provided by a tourism enterprise. It is a composite product composed of individual items and services, offering tourists a comprehensive benefit from integrating multiple elements.

3.2. *Concept of tourism marketing*

Tourism marketing is a market-driven process where enterprises place tourists at the centre and engage in analysis, planning, implementation, control, and feedback. The goal is to determine the appropriate market positioning for the enterprise's tourism products to meet the tourism market's and its participants' needs, thereby achieving a win–win outcome for tourism enterprises, tourists, and society.

Tourism marketing is the application of general marketing principles to the tourism industry. Given the characteristics of tourism, value creation for tourism enterprises or destinations lies in generating profit and favourable evaluations. For consumers, value is manifested through positive tourism experiences. Therefore, the essence of tourism marketing is to serve tourism consumers, including identifying which consumer groups to serve and determining how best to serve them.

3.3. *Elements of tourism marketing*

Both domestic and international scholars have defined tourism marketing from traditional, economic, and marketing perspectives. However, the authors propose a reverse-explanation approach, beginning with the intended direction of research and the desired outcomes, which may more accurately reflect the true nature and meaning of the tourism market.

The main elements of tourism marketing are as follows:

(1) Demand

Demand includes actual demand and potential demand. The ultimate objective of tourism marketing is to satisfy the needs of tourism consumers, emphasising likely demand. According to consumer psychology and the nature of tourism consumption behaviour, actual demand represents a conscious, analysed, and chosen need less susceptible to marketing influence. Therefore, the core objective of tourism marketing should be to identify and stimulate potential demand and convert it into actual consumption behaviour.

Since it is difficult for tourism products to satisfy both actual and potential demands simultaneously, and even more so to meet the needs of all consumers, enterprises must concentrate their efforts on one specific need within a potential consumer segment, commonly known as market segmentation and target market selection. Accordingly, enterprises should formulate targeted marketing strategies that address the distinct needs of different potential market segments.

(2) Place

"Place" includes both tangible and intangible locations. In the tourism market, tangible locations mainly refer to contractual venues, such as travel agency offices, branches, and tourist service centres, as well as consumption sites, such as scenic areas, hotels, and tourism shopping centres. In practice, much theoretical and applied marketing work has focused heavily on improving tangible locations—for instance, expanding travel agency chains and constructing large-scale service centres. However, such an approach is increasingly questionable.

The tourism market also encompasses intangible locations, particularly in the context of rapid mobile internet development. Tourists can now access information, consult prices, compare options, and purchase anytime and anywhere with mobile connectivity. This fundamental shift is gradually reshaping tourism consumption habits, methods, and the overall rules of tourism marketing. Consequently, providing more convenient digital access and service locations should be a top priority in current tourism marketing efforts.

(3) **Relationship**

"Relationship" refers to all forms of exchange relationships. In the tourism market, such relationships typically involve the right to use products or services within a specified time and area, rather than ownership exchanges (except for physical tourism goods). This is a fundamental distinction between the tourism market and other markets, and it represents the core element of tourism marketing.

A correct understanding and handling of this concept is essential. In product development and design, enterprises should prioritize enabling this form of exchange. In product packaging and promotion, emphasis should be placed on highlighting the use value of tourism products.

3.4. *Characteristics of tourism marketing*

Compared with marketing in other industries, tourism marketing possesses several distinctive characteristics, which are primarily reflected in the following aspects:

(1) **Service-Oriented**

Except for specific tangible tourism products, the essence of tourism offerings lies in providing services rather than transferring ownership. Tourists place importance not only on the quality-of-service facilities, but more significantly on the overall service experience during the tourism process—namely, the degree of enjoyment, comfort, and satisfaction derived from the service.

(2) Participatory

In the tourism market, the production and delivery of tourism products often coincide with consumer participation, both in time and space. Tourism marketing is, therefore, a dynamic process involving the periodic engagement of tourists. Maintaining active interaction with tourists has thus become a crucial component of effective tourism marketing practices.

(3) Temporal Characteristics

The tourism market exhibits strong seasonality and time sensitivity. Tourism products are inherently non-storable and non-transferable. While the facilities, equipment, and labour used to produce tourism services may exist in tangible form, they merely represent production capacity, not the tourism product itself. If a company fails to accurately seize the timing of market demand and respond promptly to consumer needs, it results in a waste of productive resources and ultimately leads to the failure of marketing efforts.

(4) Differentiation

Differentiation is an inherent attribute of tourism products, making it impossible to promote them in the same direct manner as other goods in traditional marketing fields. Instead, marketing must be integrated throughout the entire value chain, from product development and design to branding and promotion, sales operations, service delivery and support.

3.5. *Tourism marketing environment*

The tourism marketing environment refers to the ecological system composed of internal and external factors affecting a company's ability to manage its marketing activities. It includes both the macro marketing environment and the micro marketing environment.

(1) Macro Marketing Environment of the Tourism Market

The macro marketing environment consists of broader political, economic, social, and technological factors, which exert their influence indirectly. These factors affect the marketing behaviour of tourism enterprises via the microenvironment as intermediaries.

(2) Micro Marketing Environment of the Tourism Market

The micro marketing environment refers to the immediate and specific forces directly involved in a company's marketing activities.

It forms the fundamental environment for the enterprise's survival and development. These elements include the tourism enterprise, marketing intermediaries, suppliers, buyers, competitors, and public stakeholders.

Although the macro and micro environments exist in different spatial dimensions, both are indispensable in the tourism marketing process. Controllable (subjective) factors within the microenvironment cannot be isolated from the uncontrollable (objective) factors in the macroenvironment. For a tourism enterprise to achieve its marketing objectives and meet the needs of tourism consumers to the greatest extent possible, it must coordinate controllable micro factors with uncontrollable macro conditions.

This coordination must be achieved through the proactive utilization of the enterprise's marketing capacity, the strategic use of controllable elements, and a conscious adaptation to external macro conditions. The purpose of such coordination and adaptation is to satisfy target market consumers better and realise the sustainable development of the enterprise's overall tourism marketing activities.

3.6. *Tourism market segmentation*

Tourism market segmentation refers to the process by which a tourism enterprise divides the overall market into two or more subgroups of tourists based on differences in their characteristics and needs. Each subgroup formed through segmentation, characterised by similar demand patterns, is referred to as a sub-market.

The basis for segmenting the tourism market lies in the diversity of tourist needs. In the tourism industry context, such differences may manifest in various forms. According to general marketing principles, the tourism market can be segmented along four main dimensions: demographic factors, geographic factors, psychographic factors, and behavioural factors.

(1) Tourism Market Segmentation Based on Demographic Factors

The demographic characteristics of tourists can be reflected in various aspects, such as age, gender, household size, family life cycle, income, occupation, educational level, social class, ethnicity, religion, and nationality. This segmentation method is widely used, as these indicators are directly related to tourists' desires, preferences, and travel frequency, and are generally easier to measure than other variables. Therefore, demographic indicators serve as crucial bases for effective market segmentation for tourism enterprises.

(2) Tourism Market Segmentation Based on Geographic Factors

Geographic segmentation refers to dividing the tourism market based on tourists' geographic locations, enabling tourism enterprises to analyse the characteristics of various sub-markets from a regional perspective. The tourism market can be segmented into distinct sub-markets according to different geographic variables, such as region, country, city, rural area, climate, population density, and spatial distance.

The underlying rationale is that tourists from different geographic locations have divergent needs and preferences for tourism products, and they also respond differently to marketing strategies and tactics employed by tourism enterprises. Geographic segmentation typically takes the following forms:

1. **Segmentation by Region**
   The United Nations World Tourism Organization (UNWTO) divides the international tourism market into six major regions: Europe, the Americas, East Asia and the Pacific, South Asia, the Middle East, and Africa. Generally, Europe and North America generate higher volumes of outbound tourists and receive more international tourists, contributing to higher international tourism revenue. Over the past two decades, the East Asia and Pacific region has experienced the fastest growth in tourism development and international tourism income.
2. **Segmentation by Country**
   This is one of the most commonly used methods in the tourism industry. Segmenting tourists based on their country of origin helps destinations and tourism enterprises better understand key source markets, allowing them to develop tailored marketing strategies that address the specific needs and preferences of tourists from those countries, thereby improving marketing effectiveness.

3. Segmentation by Climate

Climatic conditions influence both the consumption of tourism products and the flow of tourists. In the international tourism market, tourists from colder, less sunny regions prefer to travel to warmer, sunnier destinations. This trend partly explains the thriving tourism sectors in the Mediterranean and Caribbean regions. Based on climatic characteristics, tourism markets can be segmented into tropical, subtropical, temperate, and polar zones.

4. Segmentation by Population Density

Tourism markets can also be segmented based on population density into urban, suburban, and rural areas.

(3) Tourism Market Segmentation Based on Psychographic Factors

Psychographic segmentation divides the tourism market based on psychological factors such as lifestyle, attitudes, and personality traits. Demographic factors and psychographic characteristics influence tourists' desires, needs, and purchasing behaviour. Even within the same demographic group, psychological characteristics can vary significantly. Therefore, tourism enterprises can segment the market into sub-markets based on psychographic distinctions. Common approaches to psychographic segmentation include the following:

1. Segmentation by Lifestyle

Lifestyle refers to how individuals spend their time (activities), what they consider essential (interests), and their perceptions of themselves and their environment (attitudes). It represents a pattern of living and the allocation of time and money, and it is a key factor influencing tourists' desires and needs.

Today, an increasing number of tourism enterprises segment their markets based on tourists' varying lifestyles and design corresponding products and marketing mixes tailored to each lifestyle group. Differences in lifestyle affect product design and influence pricing strategies, distribution methods, and promotional approaches. Many tourism enterprises have discovered new and highly attractive market opportunities through lifestyle segmentation.

2. **Segmentation by Attitude**
   Attitude-based segmentation refers to categorising tourists based on their attitudes toward tourism enterprises and their products and services, and then adopting corresponding marketing strategies. For example, tourists with a neutral attitude, such as "I've heard of the brand, but I don't know much about it," should be provided with detailed information and targeted through persuasive promotional campaigns. For those with a positive attitude, such as "This brand is the best in the tourism market," ongoing promotional efforts and formal engagement methods (such as contract-based partnerships) should be employed to reinforce loyalty. As for tourists with a negative attitude, such as "This brand is worse than another," it is more challenging to shift their perception. In such cases, enterprises should adopt a more meticulous promotional approach, enhance product quality, and improve brand image to influence their opinion.

3.7. *Tourism market segmentation based on behavioural factors*

Behavioural segmentation involves dividing the tourism market into different groups based on tourists' knowledge of tourism products, the benefits they seek, their consumption behaviour, or their responses to tourism products. Behavioural variables—including purchase purpose, timing of purchase, desired benefits, user status, product usage frequency, and brand loyalty—serve as key starting points for segmenting the tourism market.

1. **Segmentation by Purchase Purpose**
   According to the purpose for which tourists purchase tourism products, the market can be broadly segmented into the following categories: leisure tourism, business tourism, visiting friends and relatives (VFR), shopping tourism, working holidays, religious or spiritual exploration, adventure tourism, sports and wellness tourism, and educational tourism. These sub-markets reflect the diversity of tourist motivations and their varying demands for tourism products.
2. **Segmentation by Timing of Purchase**
   Tourism markets can also be segmented based on the timing of tourists' needs, purchases, or consumption of tourism products and services.

Some tourism offerings are specifically tailored to particular seasons or occasions, such as the Labour Day holiday, National Day, Spring Festival, or school holidays (summer and winter breaks). Tourism enterprises can utilise purchase timing as a segmentation variable, designing and providing tourism services to meet the specific needs of those timeframes.

3. **Segmentation by Desired Benefits**
   Generally, tourists purchase specific tourism products or services in pursuit of particular benefits. As such, enterprises can segment the market according to the distinct benefits sought by different tourists. When applying this method, businesses must assess which benefits are most important to tourists, identify the types of tourists seeking them, determine what benefits their products offer, and evaluate whether the benefits provided match those the target groups seek.
4. **Segmentation by User Status**
   This approach divides tourists based on their status to a given product or service, such as non-users, former users, potential users, first-time users, and regular users.
5. **Segmentation by Usage Frequency**
   Usage frequency refers to how often a tourist consumes a specific product or service. Based on this criterion, tourists may be categorised as light users, moderate users, or heavy users. Generally, moderate users are more engaged in holiday-related travel than light users; they tend to be more receptive to innovative tourism products, possess more comprehensive knowledge, and are more likely to act as opinion leaders. These tourists travel frequently and gather information through newspapers, books, and tourism expos.

3.8. *Tourism market segmentation based on tourist loyalty*

Tourist loyalty refers to the degree of persistent preference and psychological commitment a tourist exhibits towards purchasing tourism products and services from a particular brand. Based on the level of brand loyalty, tourists can be classified into the following four categories:

1. **Strongly Loyal Tourists**
   These tourists **consistently and unwaveringly purchase products and services** from the same tourism brand.

2. **Moderately Loyal Tourists**
   This group shows loyalty to **two or three tourism brands**, alternating their choices within a limited set of preferred providers.
3. **Shiftily Loyal Tourists**
   These tourists **switch their preference** from one brand over time, gradually transferring their loyalty to a different tourism provider.
4. **Brand Switchers**
   Brand switchers are tourists who **are not loyal** to any specific brand and frequently change providers without consistent preference.

Tourism enterprises can gain valuable insights by analysing tourists' brand loyalty profiles. By studying the characteristics of strongly loyal tourists, enterprises can formulate effective product and service development strategies. Companies can identify their most competitive rival brands by analysing moderately loyal tourists. By examining tourists who have shifted away from the enterprise's brand, companies can detect weaknesses in their marketing strategies and implement improvements accordingly. As for brand switchers, tourism businesses can adopt varied marketing approaches to attract and retain their interest.

4. Understanding Educational Tourism Marketing

4.1. *The concept of educational tourism marketing*

Literally, "educational tourism" combines "study" and "travel." However, a educational tourism in practice represents an organic integration of research, learning, and travel, rather than merely the sum of its parts.

In its broadest sense, a educational tourism refers to a specialised form of travel in which participants, motivated by a desire for cultural knowledge, leave their habitual residence to engage in research-oriented and exploratory learning activities in specific locations.

In a narrower sense, a educational tourism refers to an educational tourism activity primarily designed for primary and secondary school students, wherein collective travel serves as the vehicle for instruction. The purpose is to enhance students' overall competence by relying on tourist attractions and other social resources to deliver experiential and research-based learning.

Educational tourism marketing is the comprehensive process of analysing, planning, executing, and controlling all stages of a educational tourism activity, with the educational tourism participant as the target client. The aim is to develop the educational tourism market and meet the needs of its participants, while achieving mutual benefits for all stakeholders across the industry chain's upstream, midstream, and downstream segments.

4.2. *The educational tourism marketing industry chain*

The educational tourisms the marketing industry chain, which consists of upstream, midstream, and downstream sectors. The upstream segment comprises resource suppliers; the midstream comprises educational tourism enterprises; and the downstream involves schools and individual consumers.

I. **Upstream Analysis of the Educational tourism Marketing Industry Chain**
   The upstream sector of the educational tourism marketing industry chain refers to resource suppliers, which primarily include educational tourism resource suppliers, educational resource suppliers, and site resource suppliers.

   1. **Educational tourism Resource Suppliers**
      These suppliers provide essential tourism resources such as accommodation, air tickets, and attraction entry tickets. Since educational tourisms typically take place during winter and summer **holidays**, which are peak travel seasons, prices for hotels, flights, and attractions are generally higher than during off-peak periods, giving resource suppliers stronger bargaining power. Furthermore, educational tourism destinations are often located in popular cities known for rich natural and cultural resources, meaning suppliers with access to these key destination channels hold greater leverage in negotiations.
   2. **Educational Resource Suppliers**
      Educational value is a core emphasis of educational tourism products. These tours require qualified instructors to provide real-time explanations of historical and cultural content during the journey, or to lead scientific exploration and personal development activities at the camp. Such instructors are typically expected to possess advanced subject knowledge, strong professional ethics, and must undergo a rigorous vetting process before being certified to guide students.

3. **Site Resource Suppliers**
   Site suppliers provide the physical venues necessary for educational tourism operations. Since building these venues requires substantial investment and approval from government authorities due to land-use restrictions, available land resources are limited. Some educational tourism enterprises, particularly those with light assets and heavy operations, choose to purchase land and construct their own camps. During periods of non-use, they may also lease these facilities to other organisations to optimise utilisation.

II. Midstream Analysis of the Educational tourism Marketing Industry Chain

The midstream sector of the educational tourism marketing industry chain comprises educational tourism enterprises, which can broadly be categorised into two types: study-tour-based enterprises and camp-based enterprises.

1. **Study-Tour-Based Enterprises**
   This sector has attracted significant investment attention, with a growing number of participants rapidly entering the market. These enterprises include specialised educational tourism institutions, educational training organisations, and travel agencies that offer academically themed travel products.
2. **Camp-Based Educational Tourism Enterprises**
   These enterprises operate through various thematic camps, including programmes in personal development, outdoor sports, environmental education, military-style training, cultural immersion, and science and innovation. Product differentiation in outdoor survival and camp-based educational tourism products is relatively high. Factors such as camp size, the depth and engagement level of the curriculum, and the instructors' professionalism significantly influence the final pricing of these offerings.

III. Downstream Analysis of the Educational Tourism Marketing Industry Chain

The downstream segment of the educational tourism marketing industry chain consists of schools and individual consumers.

1. **Schools**
   Schools typically engage in institutional partnerships with selected educational tourism enterprises. Parents generally place greater

trust in schools, which increases their acceptance of the educational tourism providers endorsed by the schools. Given the intense competition in the sector and the minimal differentiation in product offerings, schools possess strong bargaining power, further compressing profit margins for educational tourism enterprises.

2. **Individual Consumers**

   Individual consumers are primarily students residing in first- and second-tier cities. Their parents generally possess higher levels of education, stable incomes, and substantial economic capacity. These families value holistic education, and the household environment tends to be relatively open and supportive, encouraging the development of children's interests, broadening their horizons, and enhancing their comprehensive competencies. Factors such as programme concept, safety assurance, cost-effectiveness, instructor qualifications, and itinerary design are critical in shaping consumer decisions during the purchasing process. Additionally, educational tourism enterprises can enter the market more quickly by collaborating with schools. In contrast, when targeting individual consumers directly, customer acquisition tends to be slower, although word-of-mouth marketing is often more easily achieved.

5. Chapter Summary

Through the analysis of the concepts and elements of the market and marketing, as well as the review and study of the marketing environment and strategies, this chapter has provided a deeper understanding of the characteristics of tourism marketing and its operating environment. It also explored commonly used tourism marketing strategies, laying a solid theoretical foundation for the following chapters.

6. Post-chapter Exercises

1. Analyse the similarities and differences between traditional and modern marketing philosophies.
2. Among the marketing strategies discussed in the text, which do you believe most accurately meets the needs of today's consumers, and why?

3. Summarise the key strategies used in tourism marketing and provide a relevant application case.
4. How can businesses accurately identify their target consumer groups in market positioning?
5. How can greater value be created across the educational tourism industry chain?
6. Design a creative and distinctive marketing campaign for a specific educational tourism destination.

Chapter 2

Overview of the Educational Tourism Market

Learning Objectives

1. To understand the origin and definition of educational tourisms.
2. To comprehend the historical development and current status of educational tourisms.
3. To grasp the six key characteristics of educational tourism products.
4. To master the development features of the educational tourism market.
5. To conduct a comparative analysis of the differences between educational tourism marketing and general tourism marketing.

Knowledge Framework

Overview of the Educational tourism Market

- Overview of the Educational tourism Industry.
    - Concept and development of educational tourisms.
    - Characteristics of educational tourism products.
- Overview of the Historical Development and Current Status of Educational tourisms.
- Analysis of the Development Characteristics of the Educational tourism Market.

- Current status of the educational tourism market.
- Analysis of the development characteristics of the educational tourism market.
- Challenges and solutions facing the educational tourism market.

Key Learning Points

1. The definition of educational tourisms.
2. The characteristics of educational tourism products.
3. The historical development of educational tourisms.
4. The development features of the educational tourism market.
5. The origin of the educational tourism concept.
6. The differences between educational tourism activities and general tourism activities.
7. The differences between educational tourism marketing and general tourism marketing.

Contents

1. Case Introduction

On 24 June 2021, the "Fun Sanya Family Island Adventure" 2021 Sanya Parent-Child Educational tourism Forum was held in Haitang Bay, Sanya, jointly hosted by the Sanya Tourism Promotion Bureau, organised by CTS International Convention and Exhibition (Hainan) Co., Ltd., and co-organized by China Tourism Group Hainan Regional Headquarters and other entities. Sanya Municipal Government Deputy Secretary-General Lin Xi, Deputy Director of Sanya Tourism Promotion Bureau Zhang Xiaolin, Secretary-General of the China Tourism Association Parent-Child Tourism and Youth Camp Branch Zhang Linlin, and others attended and delivered speeches. Zhang Linlin gave a keynote speech titled "*Parent-Child Tourism Destination Development—From 'Individual Tourism' to 'Small Group Tourism' Transformation*".

More than 150 tourism and educational tourism enterprises nationwide participated in this event. Publicity activities targeted the general public to promote the "Sanya's New Tourism Approach" concept. Industry experts gave insightful keynote speeches and panel discussions, such as "*New Consumer Behaviour Trends in China's Parent-Child Educational tourism Market*".

Zhang Linlin focused on the integrated development of parent-child tourism and educational tourisms, the development trends of the parent-child tourism industry, and the emerging trends in educational tourism consumption. She discussed upgrading parent-child tourism products in Sanya and various other aspects. Experts engaged in in-depth discussions and exchanges concerning the development of parent-child educational tourisms.

The "Fun Sanya Family Island Adventure" 2021 Sanya Parent-Child Educational tourism Forum served as a professional and vertical event within China, leveraging the momentum of domestic economic circulation and Sanya's strong parent-child educational tourism sector. It outlined a series of well-developed educational tourism routes and products. The "Parent-Child Educational tourism Supply and Demand Matchmaking Session" in the afternoon provided a direct cooperation platform between domestic parent-child educational tourism providers and relevant Sanya enterprises, with highly effective on-site negotiations.

As tourism enters a crucial stage of transformation and upgrading, the tourism market is becoming more segmented. The parent-child educational tourism market has exceeded one trillion yuan. Camp-style educational tourisms are gaining traction by integrating tourism, education, and residential experiences. The China Tourism Association Parent-Child Tourism and Youth Camp Branch has taken on the role of bridge and platform by cooperating with local youth organisations and associations to promote the development of China's parent-child and youth camp industry. It also fosters dialogue across sectors and regions to strengthen the tourism destination brand.[1]

1.1. *Analytical questions*

1. Who participated in the "Fun Sanya Family Island Adventure" 2021 Sanya Parent-Child Educational tourism Forum?
2. What was the purpose of the "Fun Sanya Family Island Adventure" 2021 Sanya Parent-Child Educational tourism Forum?

2. Overview of the Educational Tourism Industry

Educational tourisms are off-campus educational activities organised and arranged by education departments and schools through collective travel and group accommodation, combining investigative learning with travel experiences. In 2013, the State Council issued the Outline for National Tourism and Leisure (2013–2020), which proposed the concept of "educational tourisms" and emphasised promoting them among primary and secondary school students. In 2014, the term "educational tourisms" was officially introduced in state policy documents. In 2016, eleven ministries including the Ministry of Education jointly issued the "Opinions on Promoting Educational tourisms for Primary and Secondary School Students", and in December of the same year, the National Tourism Administration released the "Educational tourism Service Specification (LB/T 054–2016)", signifying the support from both policy and market demand that positioned educational tourisms as the "new blue ocean" of the tourism education industry.

---

[1]*Source of Case*: Official website of the China Tourism Association.

2.1. *Concept and development of educational tourisms*

1. Origin and Definition of Educational tourisms

Tourism has become an essential part of modern life and an integral part of family life. It involves travelling from one place to another, exploring the world and gaining personal insights through first-hand experiences. Travel offers broader and more complex social environments and diverse social and cultural contexts for teenagers, significantly influencing their development into adulthood (Jaleh, 2002).

Educational tourisms have been popular for centuries in the birthplace of the United Kingdom's tourism industry. From the 16th to the 18th centuries, the aristocratic "Grand Tour" served as an essential part of the education of British elites. Over time, this evolved into more widely accessible educational activities in British private schools. According to the Oxford Advanced Learner's English-Chinese Dictionary, the "Grand Tour" refers to "a long journey through Europe, undertaken in the past by rich young people from Britain to complete their education."

Economically, the Grand Tour facilitated international trade between Britain and continental Europe; culturally, it strengthened cross-cultural communication and greatly contributed to advancing British education and culture (Tao Jun, 2005). It played a pivotal role in the historical development of Britain.

In 1946, Japan had officially incorporated "school excursions" into the national education system. These excursions, defined in the Meikyo Japanese Dictionary as teacher-led group travel involving young people, represent an integral part of school life (Yang Sheng *et al.*, 2012).

In ancient China, educational tourisms—then known as "youxue" (educational tourism)—were a traditional form of learning and education. As early as the Spring and Autumn Period, educational tourism had become common, with the journey of Confucius and his disciples becoming a widely cited example. According to historical texts such as Records of the Grand Historian, educational tourism could be divided into two categories: travel to expand one's horizons, and travel organised as banquets for social status and recognition. Educational tourism has existed in China for over 2000 years and has helped address imbalances in educational development, encouraging educational exchange between regions and fostering cooperation. It is a healthy educational model based on the principle of "student-centredness", and should be supported by a relaxed political climate and a rich cultural atmosphere (Xiao Jumei & Li Rumi, 2017).

At the end of 2016, the Ministry of Culture and Tourism issued the "Educational tourism Service Specification (LB/T 054–2016)", which defined educational tourisms as "off-campus educational activities primarily for primary and secondary school students, combining group travel with educational purposes and using social resources such as tourist attractions to conduct experiential and investigative learning". This specification provided a standardised definition and practical guidance for educational tourisms, becoming a key reference for schools and related organisations.

2. The "Introduction of the Educational tourism Concept

In 2016, eleven departments, including the Ministry of Education, issued the "Opinions on Promoting Educational tourisms for Primary and Secondary School Students", which defined educational tourisms and affirmed their educational value in fostering students' socialist values, patriotism, and sense of responsibility. The policy emphasised the role of educational tourisms in enhancing the quality of education, cultivating innovative thinking, and encouraging student autonomy. It also highlighted the practical benefits for national economic development, quality of life improvement, and fulfilment of students' travel and educational needs from an early age.

The document positioned educational tourisms as an essential form of experiential education. The 2016 China Tourism Investment Report listed educational tourisms as one of the top ten sectors for future tourism investment. Despite a promising outlook, challenges remain:

- First, many educational tourism providers still conflate them with traditional sightseeing tours, lacking genuine educational content.
- Second, standards in areas such as pricing, safety, curriculum, and service vary greatly.
- Third, coordination between schools and tour operators is insufficient, and the decision-making process often lacks communication between decision-makers and the end users—students and parents.

As a result, although educational tourism products flourish, the absence of unified standards hinders healthy development. Educational institutions have begun to participate actively in this market. A collaborative and resource-integrated approach is essential to capitalise on the golden opportunity of educational tourism development.

2.2. *Characteristics of educational tourism products*

(1) Educational Attributes and Requirements of Educational tourism Products

Educational tourism products exhibit distinct educational attributes, with the principal supervisory bodies being the Ministry of Education and various levels of academic administration.

According to the *Educational tourism Service Specification* (LB/T 054–2016), issued by the China National Tourism Administration on 19 December 2016 and officially implemented on 1 May 2017, the document mandates that both organisers and hosts of educational tourism services must collaborate with schools to formulate educational service plans aligned with the foundational objectives of basic education. These plans should define the goals and content of academic activities and specify the required instructional hours tailored to different student age groups. Additionally, the specification emphasises the experiential nature of educational tourism products. Table 2.1 summarises the categorisation of service items, product classifications, and service procedures under the *Educational Tourism Service Specification*.

(2) Service Characteristics and Requirements of Educational Tourism Products

Educational tourism products fall under service-based products and possess service attributes. According to the "Educational Tourism Service Specification" (LB/T 054–2016), improving the service quality of educational tourism products is the responsibility of both the organisers and the providers. The organisers and providers are required to collect and summarise information about the quality issues of educational tourism activities promptly, clearly identify the main problems in the products, determine the specific causes of quality issues, and implement improvements through systems enhancement, enhanced training, supplier adjustments, product optimisation, service element refinement, and process improvement measures, to continuously improve the quality of educational tourism services. Specifically, improvements to educational tourism services may include:

1. Establishing a complete educational tourism service system.
2. Strengthening training for educational tourism service staff.
3. Adjusting and optimising supplier responses.

Table 2.1. Consolidation of Relevant Contents from the "Educational Tourism Service Specification".

	Category	Content
Educational tourism Product Service Items	Fitness-Oriented	Activities focused on cultivating students' survival and adaptability skills, such as hiking, challenges, camping, outdoor training, and survival/rescue exercises.
	Skill-Oriented Items	Designed to enhance students' self-management and hands-on abilities, such as integrated practical training, life experience exercises, personal organisation, and handicraft making.
	Intellect-Oriented Items	Aimed at enhancing observational and learning abilities, including site visits, lectures, recitals, and reading sessions.
	Emotion-Oriented Items	Targeted at developing emotional intelligence and implementation skills, such as moral education, team games, emotional bonding, and talent exhibitions.
Educational tourism Product Categories	**Scientific Knowledge & Popularisation Type**	Mainly includes museums, science centres, thematic exhibitions, zoos, botanical gardens, cultural heritage sites, industrial zones, research facilities, etc.
	Nature Appreciation Type:	Covers resources like mountains, rivers, lakes, seas, grasslands, and deserts.
	Field Investigation & Experience Type	Includes farms, practice bases, summer camps, or team development facilities.
	Motivational Expansion Type	Focuses on red education bases, university campuses, national defence education bases, military camps, etc.
	Cultural & Recreational Type:	Encompasses various themed parks and film studios.
Educational tourism Service Process	**Pre-Tour**	Preparatory tasks include reading relevant books, researching materials, drafting learning plans, and understanding objectives and functions.
	During Tour	Participation in educational activities, guided diary writing, or field report composition.
	Post-Tour	Reflection and sharing of experiences, e.g., essay writing contests or sharing sessions.

4. Improving the design of educational tourism products.
5. Enhancing the service elements of educational tourisms.
6. Refining the operational aspects of educational tourism services.

(3) Curriculum Characteristics and Principles of Educational Tourism Products

Educational tourism products differ from general tourism activities in that they provide curricular guidance through travel, and thus have educational characteristics.

Education exists to transmit and renew culture. For adolescents, education enables them to acquire knowledge and skills while developing personal and social attributes that help them adapt better to society. The development of educational tourism curricula allows middle and primary school students to engage in extracurricular, practical learning that enhances survival skills, life capabilities, critical thinking, and social ideals.

Educational tourism curricula development should follow three general principles:

1. **Safety First:** The safety of students participating in educational tourisms is the top priority.
2. **Interest and Enjoyment:** As extracurricular practical courses, educational tourism curricula should focus on engaging students and being enjoyable.
3. **Equitable Participation:** Every student within a group should have the opportunity to participate in and benefit from educational opportunities. Educational tourism organisations and schools must ensure that activities provide equal benefits and that underprivileged students can access subsidies or waivers.

In addition to the three principles above, curriculum development should also follow three specific guidelines:

1. **Student-Centred Learning:** Focus on active participation and stimulate curiosity and learning.
2. **Inquiry-Based Learning:** Encourage critical thinking and inspire scientific curiosity.
3. **Experience-Oriented Learning:** Ensure students participate in decision-making and task resolution through real-world practice and engagement.

Educational tourism courses may be developed in two primary ways:

- Independently through educational tourism camps or institutions
- Through collaboration between institutions and schools.

The general steps and methods for collaborative curriculum development include:

1. Drafting a course development plan based on school needs (listing course feasibility, organising entities, timelines, etc.)
2. Creating an outline for the course that progresses from simple to complex and from general to specific, featuring progressively detailed sections.
3. Coordinating student needs, school resources, and instructor participation tasks.
4. Designing the specific course content.
5. Conducting trials or test lectures of the curriculum or product, followed by revisions based on feedback.
6. Finalising and updating the curriculum version.

(4) Safety Characteristics and Management of Educational tourism Products

Educational tourism products must ensure students' travel safety. Given their safety characteristics, there is a need for strict safety management throughout the activity. Educational tourism service providers, including organisers, contractors, and responders, must establish and implement a sound safety management system and related contingency plans (especially for emergencies such as natural disasters, fires, food hygiene, and public security incidents). They must also regularly conduct drills and establish a robust and efficient safety risk control mechanism with a "trinity" structure to ensure the smooth and safe execution of educational tourism activities.

The formulation of safety management systems for educational tourism services should include the following: safety management procedures for educational tourisms, operational guidelines for emergency plans, product safety assessment systems, training plans for tour guides and instructors, and protocols for training safety management personnel.

A dedicated safety education and training programme should be developed for service providers' safety management personnel. This includes the following: safety operation procedures, responsibilities and requirements, emergency handling procedures, etc.

Service providers should offer various types of safety education for students participating in the educational tourisms, including basic safety reading materials. Pre-departure meetings or briefings should be conducted to raise safety awareness and reinforce the students' safety precautions.

(5) Responsibility Characteristics and Complaint Handling of Educational Tourism Products

A high level of responsibility characterises educational tourisms. Before the start of the educational tourism, both the organiser and contractor should clarify their respective responsibilities. Should issues arise during the educational tourism, following the "Educational tourism Service Specification" (LB/T 054–2016), contractors must take full responsibility for complaint handling and related matters. They must announce complaint hotlines, complaint handling procedures, and response timelines in advance. Additionally, they must establish complaint records and feedback mechanisms. Therefore, throughout the activity, responsibility for complaint resolution must rest entirely with the contractor.

(6) Experiential Characteristics and Activity Handbook of Educational Tourism Products

Educational tourism products are educational offerings centred on student experience and necessitate the design of an experience-focused curriculum. Institutions creating these products must also develop an activity handbook for students. The handbook should take into account the following aspects:

1. **Course Introduction** (including cover design, concept, characteristics, and relevance to the school's basic curriculum).
2. **Specific Course Objectives**
    - Objective 1: Knowledge acquisition.
    - Objective 2: Skill development.
    - Objective 3: Emotional goals.

3. **Course Content** (detailed outline and teaching methods).
4. **Instructor Profile** and Experience.
5. **Note-taking or drawing sections**.
6. **Evaluation Methods**, including timing and assessment criteria.
7. **References** and platforms used during the activity.
8. **Post-course Summaries and Feedback**, including student, teacher, and programme evaluations.
9. **Additional Information** such as preparation checklist, attention points, course leaders' contact information, etc.

3. An Overview of the Historical Development and Current Status of Educational Tourisms

Educational tourisms refer to a form of educational activity carried out through travel. In today's globalised world, short-term travel experiences appeal to students from various academic backgrounds (Mikoski, 2013). Although educational tourisms are a recognised part of the educational journey, limited research remains on their educational value (Otero-Urtaza, 2012), and their connection to education continues to be an under-explored area (Falk *et al.*, 2012). Despite a growing awareness over the past two decades, the educational role of travel has only recently become a serious subject of academic inquiry and recognition.

During the Spring and Autumn period of Chinese history, a system resembling teacher-led student travel for educational purposes had already taken shape. For example, Confucius led his disciples to travel and study across different states. In classical Chinese texts, phrases such as "carry books while travelling" and "study through travel" are frequently seen. According to "Educational tourism Chronicle", educational tourisms were divided into two main categories: one aimed at enhancing students' understanding and perspectives through exploration, and the other focused on participation in diplomatic or elite banquets to gain practical exposure. Educational tourisms were considered an educational activity under "learning through experience". They were seen as fostering individual development and moral character in a relaxed yet artistically rich political environment (Xiang Ju Mei & Li Ru Xuan, 2017).

The tradition of educational tourisms (the "Grand Tour") as educational tourism has existed for centuries in Britain. From the 16th to the 18th centuries, the Grand Tour was viewed as a finishing stage in the education of aristocrats (French *et al.*, 2000), and it became a key component of aristocratic education. Early educational tourism focused on personal development through immersive experiences rather than fixed tourist sites (Towner, 1996). In Europe, this kind of travel was initially limited to the aristocracy (Towner, 1985), and it wasn't until the 19th and 20th centuries that such educational tourism was made accessible to the broader public.

In 1882, Cossio conducted a study in Switzerland that sparked discussions on the importance of public education in Spain. Upon returning, he proposed public education reform, including establishing school trips as a fundamental teacher training component. This laid the foundation for future improvements to Spanish educational policy and impacted educational philosophy (Otero-Urtaza, 2012).

20th Century to Present: Integration of Education and Tourism

By the 20th century, the divide between work and leisure became more distinct (Roberts, 1999); yet modern tourism began integrating travel with learning and consumer experiences (Urry, 2005). Tourism became increasingly multi-functional, connecting with a variety of sectors. Educational tourism, of course, stood out as a key development. In the 1920s and 1940s, Japan institutionalised "educational tourisms" into their formal educational system. These trips were led by teachers and framed as an essential part of holistic student development, with cultural classics often being taught as part of the travel curriculum (Yang Zhi *et al.*, 2012). These trips enabled students to understand real-life contexts, expand their worldview, and enhance civic and cultural competencies.

In ancient China, even before Confucius, people frequently travelled for learning. During the 17th to 18th centuries, the "Grand Tour" became a formalised form of aristocratic travel. The 19th century saw "educational excursions" gain popularity among Western intellectuals. By the 20th century, countries like the UK had begun systematically incorporating study travel into public education systems, while China and other nations followed suit. Tourism began shifting from satisfying individual desires to fulfilling public educational objectives.

As mass tourism grew, so did various forms of tourism-related educational activities. In the early 20th century, studies by Smith and Jenner (1997) indicated that educational tourism was closely tied to national pride and global citizenship. Later, as the travel industry evolved (Burkart & Medlik, 1981), educational tourismism came to be recognised as a key component of public education (Kalinowski & Weiler, 1992). In recent decades, this form of experiential learning has increasingly been seen as an essential part of middle school education in China and internationally.

4. Analysis of Educational Tourism Market Development Characteristics

4.1. *Current development of the educational tourism market*

Since 19 December 2016, when eleven departments, including the Ministry of Education, jointly issued the Opinions on Promoting Educational tourisms for Primary and Secondary School Students, educational tourisms have been incorporated into the national comprehensive practical curriculum system. This has made them a key component of experiential education and an essential pathway for fostering innovation in education and integrating extracurricular learning with practical education. With strong government support, the educational tourism market has developed rapidly.

According to statistics, in 2017, the number of educational tourism participants in China reached approximately 3.55 million, with 2.75 million domestic participants and 800,000 outbound participants. At the time, the recorded number of educational tourism participants was still separate from the broader category of tourism statistics.

In 2018, the *White Paper on the Development of Educational Tourisms in China 2019* estimated that the domestic educational tourism market had reached a scale of 12.5 billion RMB, with participants exceeding 31.17 million.

Educational tourisms sit at the intersection of "tourism", "education", and "culture". According to multiple research reports, China's educational tourism market is projected to surpass 100 billion RMB in the coming years. But as this rapidly growing new market expands, what opportunities and challenges lie ahead? What future trends can we anticipate?

4.2. *Key features of educational tourism market development*

1. Huge Market Scale

The White Paper on the Development of Educational tourisms in China 2019 indicates that school participation in educational tourisms has increased significantly in recent years. The total market size is approaching 100 billion RMB, establishing a new "blue ocean" in the tourism industry. Educational tourisms now encompass multiple themes, including historical and cultural experiences, aerospace science, outdoor survival, and patriotic education, with emerging new formats and topics.

In 2017, the number of primary and secondary school students in China exceeded 261 million (source: *Statistical Communiqué of the People's Republic of China on the 2017 National Economic and Social Development*). Naturally, such a large demographic base has attracted significant market attention.

2. Intense Competition

The rapid growth in consumer demand for educational tourisms has led to swift market expansion. However, due to relatively low entry barriers and high market expectations, the overall quality of supply levels is uneven. According to the White Paper on the Development of Educational tourisms in China 2019, the market is expected to see explosive growth over the next 3–5 years. Nevertheless, many enterprises entering the field are still navigating early-stage competition and brand building.

Between 2010 and 2019, educational tourism products diversified significantly. In 2019 alone, the number of registered educational tourism enterprises exceeded 60,000—a 169% year-on-year increase.

Educational institutions, travel agencies, training institutions, and overseas study services have evolved into educational tourism providers. However, many providers still emphasise tourism as the primary focus and have not fully grasped the integration of education and travel. There is an urgent need for enhanced regulation and structural improvement.

3. Competitive Pricing Advantage

Educational tourism products offer a notable price advantage compared to traditional tourism products. The *White Paper on the Development of Educational Tourisms in China 2019* notes that parents are generally willing to pay a reasonable fee for educational tourisms with guaranteed quality.

For short- and medium-duration educational tourisms targeting primary and secondary school students:

- Most parents are willing to pay **3,000–5,000 RMB** per trip.
- Some are willing to pay **5,000–10,000 RMB**.
- Overseas educational tourisms can reach **12,000 RMB** per trip.

4. Shortage of Industry Talent and Inadequate Training

Most frontline personnel in the educational tourism industry lack rigorous professional training. On 18 October 2019, the Ministry of Education announced that the discipline of "Educational tourism Management and Service" would be added to the list of vocational majors, with enrolment starting in 2020. These primary aims are to meet the widespread demand from schools at all levels for organising educational tourism activities and to serve as a critical platform for cultivating national-level talent for the educational tourism industry.

In the face of this widespread talent shortage, relevant educational tourism organisations—aside from hiring experienced experts—should also consider conducting systematic training and building a comprehensive talent development system incorporating tourism, education, and practical knowledge. Training should aim to produce compound talents who combine tourism and educational expertise and can guide and manage student groups effectively. This approach will help improve educational tourism providers' service quality and professionalism.

5. Weak Infrastructure and Facility Development

Most educational tourism institutions do not have fixed facilities or dedicated activity sites. Due to the high cost of building infrastructure for educational tourisms, many organisations lack the resources to build or invest in such facilities independently. Instead, they partner with schools or rent spaces to carry out their activities.

In addition, some organisations attempt to educational tourisms by leasing or sharing facilities without long-term planning or customisation, impacting their offerings' continuity and quality.

4.3. *Challenges and solutions in the educational tourism market*

Due to the lack of unified industry entry standards, the current development of the educational tourism market across regions is uneven. Problems such as high pricing, lack of standards, inconsistent quality, unprofessional tour instructors, and absence of integrated service platforms remain prominent. To resolve these issues, the following areas should be addressed:

1. Government Involvement in Top-Level Design and Guidance

Governments should establish relevant legal frameworks, improve regulatory systems, and clearly define the responsibilities of relevant departments to support the healthy development of the educational tourism market. From a long-term perspective, in addition to improving evaluation standards for tourist sites and destinations and establishing national and local benchmarks, the government should provide special development funds for educational tourism projects and promote the overall improvement of market quality and standards.

2. Resource Sharing and Collaborative Growth

Currently, many regions operate in isolation. Breaking through geographical limitations and establishing authoritative industry associations is necessary to enhance stakeholder communication and coordination. Leveraging the Internet and network technologies can facilitate better resource-sharing platforms.

3. Development and Supply of Specialised Talent

Culturing specialised talent for the educational tourism industry must be encouraged by establishing related majors, such as education-based tourism programmes at higher education institutions. Beyond curriculum setup, it is also important to improve salaries and career progression paths for professionals to enhance recruitment and long-term development.

4. Development of Professional Educational Tourism Bases (Camps)

Unlike traditional tourism, educational tourism bases (or camps) focus more on educational attributes. Educational tourism enterprises should follow the standards and policies set by national and provincial authorities to create innovative products that realise the value of professional educational tourism bases.

Professional educational tourism bases (camps) should be safe and equipped with appropriate facilities and space for course development, experiential zones, and capable operational teams and instructors. Due to the complexity of managing and operating such bases and high investment and operational costs, substantial financial backing is essential.

5. Chapter Summary

This chapter analysed the origin and definition of educational tourisms, explained the six key characteristics of educational tourism products, examined the historical evolution and current development trends of educational tourisms in China, and compared the educational tourism market with the traditional tourism market, offering insights for future development.

6. Post-chapter Exercises

1. What is a educational tourism?
2. What are the six main characteristics of educational tourism products?
3. Why has the educational tourism industry become a hot topic in domestic tourism?

Chapter 3

Educational Tourism Market Environment Analysis

Learning Objectives

1. Understand the current state of the educational tourism market environment.
2. Understand the macroeconomic environment of China.
3. Understand the microenvironment of educational tourism marketing.

Knowledge Framework

- **Educational Tourism Market Environment Analysis**
    - **Macroenvironment Analysis of Educational Tourism Marketing**
        - Educational tourisms and the macroeconomic environment.
        - Educational tourisms and the industry political environment.
        - The socio-cultural, technological, and natural ecological environments of educational tourisms.
    - **Microenvironment Analysis of Educational Tourism Marketing**
        - Companies, suppliers and marketing intermediaries.
        - Educational tourism consumer groups.
        - Educational tourism industry competitors.
        - The general public.

Learning Focus

1. Analysis of the political environment of the educational tourism industry.
2. Analysis of educational tourism target groups.
3. Analysis of the macroenvironment of the educational tourism market.
4. Analysis of the microenvironment of educational tourism marketing.

Contents

1. Case Introduction

China's GDP in 2019 neared 100 trillion yuan, increasing by 6.1%. Per capita GDP surpassed $10,000—remarkable.

Another significant leap!

On 17 January 2020, China's economic data for 2019 was enthusiastically released. According to preliminary calculations, China's gross domestic product (GDP) for 2019 was 99.0865 trillion yuan, a 6.1% increase from the previous year; calculated at the average annual exchange rate, per capita GDP exceeded $10,000, reaching $10,276.

China's comprehensive national strength has increased, and the "family wealth" has grown even more solid.

In 2019, China's GDP was close to 100 trillion yuan. Based on the average annual exchange rate, it equated to $14.4 trillion, securing its place as the second largest economy in the world. Looking back at history, accumulating such a rich economic foundation was no easy feat.

In 1952, China's GDP was only 67.9 billion yuan. In 1986, the economic aggregate exceeded 1 trillion yuan. It took 37 years to grow from a war-torn, underdeveloped country to an economy of 1 trillion yuan. It took 14 years for 1 trillion to surpass 10 trillion yuan in 2000, which took 14 years. The transition from 10 trillion to nearly 100 trillion yuan in 2019 took only 19 years.

"In terms of per capita figures, China's developing country status has not fundamentally changed", said Wang Yuanhong. Regarding investment levels, China's per capita infrastructure capital stock is only 20%–30% of that in developed countries. Regarding consumption structure, China's 2019 Engel coefficient was 28.2%, still higher than that of developed countries, indicating that Chinese households still spend a large portion of their income on basic needs such as food and clothing.

Xu Wei also pointed out that there is still a gap between China and the threshold for high-income countries, and it is necessary to focus on promoting high-quality development.

"Income determines consumption. The continuous increase in disposable income means continued expansion in consumption scale, ongoing upgrading of consumption, and continuous optimisation of the economic structure", Wang Yuanhong said.

In 2019, the role of consumption as the "main engine" of economic growth was further consolidated. Final consumption expenditure contributed 57.8% to GDP growth, 26.6 percentage points higher than the contribution of gross capital formation. Of total per capita consumption expenditure nationwide, service consumption accounted for 45.9%, a 1.7 percentage point increase from the previous year.[1]

1.1. *Reflective questions*

1. Has China already entered the ranks of developed countries?
2. How does China's macroeconomic development affect the educational tourism market?

---

[1]*Source*: *People's Daily*, 18 January 2020.

2. The Macro Environment of Educational Tourism Marketing

The macro-environment of educational tourism marketing encompasses several key dimensions: the broader economic environment, the political and regulatory framework of the industry, socio-cultural factors, scientific and technological development, and the natural ecological context.

2.1. *Educational tourisms and the macro-economic environment*

1. Rapid Economic Growth and Shifting Consumption Patterns in China

According to official data from the Chinese government, China's GDP approached 100 trillion yuan in 2019, with an annual growth rate of 6.1%. Per capita GDP exceeded USD 10,000, placing China firmly among the world's middle-income countries. The country's Human Development Index (HDI) also continues to rise steadily.

From a consumption perspective, the Engel coefficient—a measure of household expenditure on food—stood at 28.2% in 2019, still higher than that of developed economies. This indicates that a significant portion of household income in China is still allocated to basic needs such as food and clothing. It also highlights the gap between China and high-income nations, underscoring the need to advance high-quality development across all sectors.

The *Statistical Communiqué on the 2019 National Economic and Social Development of the People's Republic of China* reported that the country's GDP reached 99.0865 trillion yuan in 2019, a 6.1% increase from 2018. Between 2015 and 2019, annual GDP growth consistently exceeded 6%. In 2019, per capita GDP rose to 70,892 yuan, reflecting a 5.7% increase year-on-year. Overall, China's economy remained stable, steadily advancing to higher levels of development, with improved quality of growth, enhanced living standards, social prosperity, and notable improvements in environmental sustainability. The country is making solid progress toward building a moderately prosperous society in all respects.

2. Changes in Household Consumption Patterns

a. **Steady Growth in Disposable Income and Consumer Spending**
   The same statistical communiqué noted that in 2019, per capita disposable income in China reached 30,733 yuan, an 8.9%

increase from the previous year. After adjusting for inflation, the real growth rate was 5.8%. Meanwhile, per capita consumer expenditure rose to 21,559 yuan, up 8.6% from 2018, with a real growth rate of 5.5%. Education, culture, and entertainment spending accounted for 11.7% of total household consumption.

Despite the onset of the COVID-19 pandemic in 2020, China's GDP grew by 2.3% compared to 2019. Per capita disposable income, adjusted for inflation, rose by 2.1%, though real per capita consumption expenditure fell by 4.0%.

b. **Tourism Activity Strongly Influenced by Economic and Social Factors**
According to the 2019 statistical communiqué, key tourism indicators—including the number of domestic trips, inbound visits, overnight stays, international tourism revenue, and outbound travel—showed upward trends. However, in 2020, due to the COVID-19 pandemic, the number of domestic trips dropped by 52.1% compared to 2019, with other tourism metrics declining by approximately 50%. From 2021 to 2022, tourism figures exhibited a noticeable rebound.

3. **Educational tourisms at a Critical Stage of Development**
China is currently undergoing its **third phase of upgrading its consumption structure, which is** characterised by a rapid rise in spending on tourism, leisure, and education. Education, culture, and entertainment expenditures have surpassed **10% of per capita consumer spending** nationwide. As a popular form of service-oriented consumption, **educational tourisms** have entered a **significant window of development opportunity**.

2.2. *Educational tourisms and the industry's political environment*

National and local government departments have recently issued policy guidelines concerning the educational tourism sector. Coupled with strong market demand and the benefits of a large population base, the educational tourism industry has entered a period of **rapid expansion**.

The market is susceptible to shifts in the political environment. As Kotler *et al.* (2002) noted, the political environment consists of laws, governmental institutions, and pressure groups that influence and regulate organisations and individuals in society. This section focuses on changes in political trends, especially the impact of **new moral education policies** on the development of the educational tourism market.

1. **New Moral Education Policy**

   a. **Overview of the New Policy**
   To improve the moral education of primary and secondary students, the **Ministry of Education of the People's Republic of China** issued a normative policy document in August 2017 titled *Guidelines on Moral Education in Primary and Secondary Schools*. This document applies to all regular primary and secondary schools across the country.
   Regarding guiding principles, the Guidelines emphasise a **student-centred approach**, placing moral education as a top priority. They aim to cultivate and implement the core socialist values, focusing on developing students' moral character and sound personality. The core goal is to nurture good behaviour and habits through a well-rounded moral education framework, integrating academic learning with practical activities such as productive labour and social engagement. It also calls for synergy between **school education**, **family education**, and **community-based learning**, promoting the long-term effectiveness of moral education in schools nationwide.

   b. **Core Principles of the Guidelines**
   The Guidelines are founded on four core principles:

First, maintaining the correct direction ensures that moral and political education in schools remains aligned with socialist educational goals and preserves the ideological leadership of schools in this domain.

Second, it is essential to respect developmental laws and align them with students' age characteristics, cognitive development, and educational patterns. The policy encourages coherence across school stages, the unity of knowledge and action, and a stronger emphasis on moral practice, emotional development, and behavioural habits. The aim is to make moral education more engaging, impactful, relevant, and effective.

Third, promoting coordinated collaboration—reinforcing the leading role of schools while mobilising families and communities to recognise and share responsibility in student development. This collective effort seeks to enhance support for students' growth and well-being.

Fourth, institutionalising regular implementation—embedding moral education into the routine functioning of schools through institutionalisation and innovation. The Guidelines advocate for integrating moral education into all aspects of school life, thereby establishing a **sustainable, long-term mechanism**.

These principles, laid out in the 2017 *Guidelines on Moral Education*, serve as a national directive for primary and secondary school moral education. During implementation, educational authorities at all levels, schools, school Communist Party organisations, and principals are expected to treat moral education as a core responsibility and strengthen efforts in the following areas:

(1) Enhancing leadership and organisational capacity in moral education.
(2) Strengthening institutional and resource support for moral education.
(3) Strengthening the development of moral education personnel.
(4) Enhancing supervision and evaluation mechanisms in moral education.
(5) Promoting scientific research in moral education.

For effective implementation, moral education must align with key aspects such as stage-specific learning objectives, content design, practical approaches, and performance standards. Specifically, it should meet the following criteria:

(1) It must be consistent with the goals and guiding principles outlined in the *Guidelines on Moral Education in Primary and Secondary Schools*.
(2) It must reflect the school's local context and instructional goals.
(3) It must cater to students' physical and psychological developmental characteristics across different age groups.
(4) It must support the holistic improvement of students' competencies.

According to the Guidelines, the goals, content, practical pathways, and requirements of moral education across different school stages are summarised in Tables 3.1–3.3, respectively.

Table 3.1. Guidelines.

School Stage	Character Development	Knowledge Development	Emotional and Behavioural Development
Lower Primary	Develop honesty, integrity, and kindness; instil modesty, diligence, and a positive mindset with good moral qualities.	Understand basic knowledge about nature, society, and the nation; develop awareness of environmental protection and respect for resources.	Cultivate love for the Communist Party of China (CPC), the motherland, and the people; develop empathy and care for others; foster basic civil behaviour and habits.
Upper Primary	Develop honesty, trustworthiness, friendship, tolerance, self-respect, optimism, and other good character traits.	Understand social changes and the nation's history; appreciate traditional Chinese culture and revolutionary heritage; learn daily life etiquette and social norms; develop initial awareness of citizenship and the rule of law.	Deepen love for the CPC, the motherland, and the people; develop civic habits and behaviours; enhance awareness of environmental protection and civic responsibility.
Lower Secondary	Cultivate respect for others, teamwork, helpfulness, and a pioneering spirit.	Understand basic social norms and moral standards; develop a sense of responsibility, the rule of law, civic awareness, and learn pathways for psychological well-being and healthy development.	Strengthen love for the CPC, the motherland, and the people; inherit and promote traditional culture and socialist values; foster self-awareness and the ability to make moral judgments.
Upper Secondary	Cultivate independence, self-confidence, and a strong sense of values and life direction.	Develop a spirit of inquiry and critical learning; understand the rule of law, Marxist beliefs, and pathways for personal growth.	Deepen love for the CPC, the motherland, and the people; uphold core socialist values with Chinese characteristics; enhance moral reasoning, self-awareness, and social responsibility; form sound civic, legal, and political understanding; learn to analyse and resolve ethical dilemmas.

Table 3.2. Moral education content in primary and secondary schools.

Category	Knowledge Objectives	Competency Objectives	Affective Objectives
Ideological and Political Education	Promote the study of Marxism and Mao Zedong Thought; strengthen learning in the theoretical system of socialism with Chinese characteristics; deepen understanding of Xi Jinping Thought through key speeches and strategies; reinforce patriotic education and revolutionary Traditions; enhance education on modern history, national conditions, and civic values.	Deepen understanding of the CPC's history, the history of New China, reform and opening-up, and socialism's development; cultivate ideals and beliefs rooted in socialism with Chinese characteristics and the Chinese Dream.	Foster political identity and emotional commitment to the Party, the nation, and its values; enhance ideological confidence and the conviction to contribute to the common goals of socialism with Chinese characteristics.
Education on Core Socialist Values	Deepen understanding of national education, patriotic education, legal education, honesty education, and civility education.	National level: Develop awareness of prosperity, democracy, civility, and harmony; Social level: Foster values of freedom, equality, justice, and the rule of law; Individual level: Instil values such as patriotism, dedication, integrity, and friendship.	Integrate core socialist values into daily education, cultivating clear moral awareness and internalised ethical standards.

(*Continued*)

Table 3.2. (*Continued*)

Category	Knowledge Objectives	Competency Objectives	Affective Objectives
Education on Core Socialist Value	Deepen understanding of national education, patriotic education, legal education, honesty education, and civility education.	National level: Develop awareness of prosperity, democracy, civility, and harmony; Social level: Foster values of freedom, equality, justice, and the rule of law; Individual level: Instil values such as patriotism, dedication, integrity, and friendship.	Integrate core socialist values into daily education, cultivating clear moral awareness and internalised ethical standards.
Traditional Chinese Culture Education	Conduct education on family love and patriotism, social responsibility, and personal development.	Learn about the historical heritage and spiritual richness of traditional Chinese culture.	Strengthen cultural identity and confidence.
Ecological Civilisation Education	Strengthen education on thrift and environmental protection; provide basic education on land, water, and food resources; promote water conservation and eco-friendly practices.	Understand the geography and natural landscape of the nation; implement behaviours such as rubbish sorting, energy conservation, and loving nature; foster ecological awareness and protection.	Cultivate habits of environmental responsibility, self-discipline, and sustainable living.
Mental Health Education	Promote self-awareness, respect for life, emotional regulation, interpersonal relationships, and social adaptation; educate students on planning for life and setting goals.	Develop mental resilience, autonomy, adaptability, and coping strategies.	Foster a healthy personality and cultivate positive psychological qualities and attitudes.

Table 3.3. Practical pathways and requirements of moral education in primary and secondary schools.

Practical Pathway	Standards for Implementation Within the Curriculum	Method: Moral Education Practice Within the Curriculum
Pathway 1: Curriculum Integration		
① Clarify each subject's moral educational aims, integrating moral education's goals into refining and implementing subject-based teaching. ② Promote rule-of-law education, anti-corruption education, civics education, etiquette education, environmental education, mental health education, labour education, anti-cult education, media literacy education, and thematic education.	① Follow the principles of curriculum-based moral education in general education, ensuring straightforward integration of moral goals into lesson planning and classroom activities, avoiding a mere tokenistic approach. ② Ground curriculum objectives in students' daily lives and real-world scenarios. Draw upon relevant political, cultural, and social resources to design course content and foster experiential learning; focus on emotional development, values, and moral decision-making.	① In language, history, and geography subjects, guide students to think about values through stories, literary texts, and discussions on human experiences. ② In science subjects such as physics, chemistry, and biology, cultivate students' scientific spirit, critical thinking, and research competence, nurturing a desire for truth and innovation. ③ In music, physical education, art, and other aesthetic subjects, enhance students' aesthetic awareness, physical well-being, and appreciation of beauty and morality. ④ In foreign language courses, broaden students' global outlook and intercultural understanding, cultivating values of openness, cooperation, and mutual respect. ⑤ Labour education should strengthen students' work ethic, develop basic labour competencies and positive attitudes toward practical tasks.

(Continued)

Table 3.3. *(Continued)*

Practical Pathway	Standards for Implementation Within the Curriculum	Method: Moral Education Practice Within the Curriculum
		⑥ Moral education should be integrated into local courses by drawing upon regional history, ethnic customs, traditional culture, major historical events, and local development practices. Such integration guides students to appreciate the richness of local culture, cultivate emotional connections to their communities, and enhance environmental and civic awareness.
Pathway 2: Cultural Education		
Utilise local conditions to improve school culture, ensuring order, cleanliness, and an aesthetically pleasing environment. Foster a positive and unique campus culture to raise the overall level of moral civilisation, making the campus a place where students feel proud to belong.	Cultivate a unique school culture; standardise campus environmental aesthetics, civility activities, and network culture.	① Beautify and optimise the school environment. ② Design appropriate school symbols, such as uniforms, slogans, and routines. ③ Organise themed civility-related school culture events. ④ Create a clean and uplifting online culture within the school.
Pathway 3: Experiential and Activity-Based Education		
① Carefully design and organise themed activities with clear topics, rich content, and diverse formats to ensure substantial educational impact.	① Regularly organise and conduct school-wide activities.	① Organise traditional festival celebrations to promote awareness of cultural traditions (e.g. Lunar New Year, Tomb Sweeping Festival, Mid-Autumn Festival, and Dragon Boat Festival).

② Help students develop good ideals, beliefs, and behavioural habits.	② Create implementation plans for various categories of student activities. ③ Conduct activities according to category-specific schedules. ④ Develop standardised formats for themed activities and enhance their ceremonial aspects. ⑤ Improve student club and activity management systems; establish a positive student club culture and ensure time and space are allocated for club participation.	② Implement seasonal or thematic activities (e.g. tree planting, poetry recitals, Children's Day, community outreach). ③ Organise commemorative activities based on key dates (e.g. National Constitution Day, World Health Day) or moral/civic education themes. ④ Launch innovative and diverse campus activities, strengthening civic spirit, moral understanding, and collective identity. ⑤ Promote sports, arts, and science activities to encourage balanced physical and mental development.
Pathway 4: Social Practice Education		
Organise themed community service and experiential activities, allowing at least one week per term. ① Traditional culture education. ② Revolutionary tradition education. ③ Rule of law education. ④ Arts and culture education. ⑤ Science education. ⑥ National Defence education.	① Tailor experiential activities to the school's philosophy and the developmental characteristics of students at different ages; align activities with relevant curriculum content to ensure moral and civic significance.	Use external venues and platforms for theme-based moral education or off-site learning. Activities can be planned by topic or grade level: 1. Historical and cultural venues: museums, memorial halls, cultural parks. 2. Civic institutions: courts, police stations, public service halls. 3. Arts and humanities: galleries, theatres, concert halls.

(*Continued*)

Table 3.3. *(Continued)*

Practical Pathway	Standards for Implementation Within the Curriculum	Method: Moral Education Practice Within the Curriculum
⑦ Environmental protection education. ⑧ Safety education. ⑨ Education on special groups (orphans, people with disabilities, etc.). ⑩ Health education.	② Organise educational tourisms and design corresponding educational and teaching plans. ③ Maintain student volunteer service records and establish personal service portfolios.	4. Science and innovation: labs, innovation centres, tech parks. 5. National defence education: military museums, defence centres. 6. Environmental protection: recycling facilities, energy and pollution treatment plants. a. Safety education: fire stations, emergency rescue centres, disaster preparedness sites. b. Social care: orphanages, homes for people with disabilities, child welfare centres. c. Health education: sports science institutes, mental health clinics, children's wellness centres.
Pathway 5: Governance-Oriented Education		
① Promote modernisation in school governance. ② Improve school governance efficiency. ③ Refine every detail of the moral education system through sound school management mechanisms.	① Improve the school's management system. ② Establish democratic class-level governance structures to promote student self-management and participatory decision-making.	① Formulate reasonable school rules and disciplinary regulations; improve the school governance system by standardising administrative behaviour. Establish a code of conduct that is recognised and consciously followed by all staff and students. ② Develop clear and reasonable class regulations, encouraging student self-discipline and democratic management.

	③ Develop systems to prevent and address bullying and school violence; establish emergency response protocols, including early warning mechanisms, intervention measures, and post-incident support. ④ Collaborate with relevant authorities to implement comprehensive safety management in the areas surrounding the school; strictly penalise unlawful behaviours that harm students' physical and mental well-being. ⑤ Clarify and standardise regulations governing teaching and administrative staff. ⑥ Strengthen the professional ethics and conduct of teachers. ⑦ Refine codes of conduct for student behaviour. ⑧ Show care and provide support for special-needs and vulnerable student groups.	③ Conduct educational programmes to raise awareness about bullying; establish a comprehensive emergency response plan. ④ Cooperate with local authorities to eliminate unlawful acts that harm students' physical and mental well-being. ⑤ Establish detailed rules and regulations for all school personnel; clarify the educational responsibilities of each position and standardise staff conduct. ⑥ Implement the *Code of Conduct for Primary and Secondary School Students (2015 Revision)*, and, in line with the school's actual circumstances, formulate and conscientiously enforce appropriate behavioural standards. ⑦ Improve systems for communication and care within the school; provide targeted support for students and staff with special needs.

(*Continued*)

Table 3.3. (*Continued*)

Practical Pathway	Standards for Implementation Within the Curriculum	Method: Moral Education Practice Within the Curriculum
Pathway 6: Collaborative Education		
Collaborate with parents and the wider community, and educational tourism organisations to foster moral development.	① Establish a sound family education system. ② Establish a comprehensive mechanism for societal co-education. ③ Establish a collaborative framework for working with educational tourism organisations.	① Actively engage with parents and encourage their involvement in school moral education efforts. ② Strengthen attention to home–school collaboration; involve families, religious institutions, and communities in building supportive educational environments. ③ Use parent committees, parent councils, home–school forums, and similar platforms to facilitate communication. ④ Proactively engage with government departments and link to resources from industry experts, specialists, and respected figures such as retired educators to provide mentorship. ⑤ Establish partnerships with public service organisations to share resources and promote the holistic development of primary and secondary school students. ⑥ Establish close cooperation with educational tourism agencies and tailor practical learning activities based on local characteristics and students' individual needs.

c. Overall Objectives of Moral Education in Primary and Secondary Schools

The overarching goal of moral education in primary and secondary schools is to thoroughly implement the fundamental task of cultivating virtue and nurturing character (立德树人). This involves enhancing the guidance for moral education work, ensuring that the Party's and the State's requirements are effectively detailed, internalised, and implemented. The aim is to construct a well-structured moral education system that is directionally sound, content-rich, developmentally coherent, diverse in delivery, and integrated into everyday school life. Furthermore, efforts should be made to professionalise, standardise, and enhance the effectiveness of moral education, thereby fostering a comprehensive framework where **all staff are responsible for education**, education occurs **throughout the student's school journey**, and ethical development is addressed in **all dimensions of student life**.

3. The Future Direction of New Moral Education

Moral education is a global priority and a growing trend in education. It is a product designed to serve a country's political, economic, and cultural systems, and domestic and international contexts have shown considerable emphasis on its development (Ren Yi, 2016). As a social construct, moral education varies across countries and periods, each shaped by distinct socio-political norms and values.

From a disciplinary perspective, moral education requires an integration of theory and practice. Upholding the correct political orientation is its foundational principle, while practice is the key measure of its effectiveness and the level of attention it receives. Establishing a supportive moral education network and cultivating a positive environment are essential for successful implementation. Strengthening the role of the school's Party organisation serves as the fundamental safeguard for delivering moral education (Li Changxi & Zhuo Xiaojun, 1992).

From a functional standpoint, moral education is an instrument that supports a nation's political, economic, and cultural systems. However, it faces domestic and international challenges (Ren Yi, 2016). According to Professor Tan Chuanbao, a leading Chinese scholar in moral education, the public often misinterprets its purpose. He identifies several intrinsic limitations:

- **First**, the function of moral education is **limited in scope**; it cannot fulfil all societal expectations.

- **Second**, the **impact on students varies by individual**, reflecting differences in receptiveness and learning conditions.
- **Third**, moral education differs in effectiveness and influence depending on the targeted moral quality and delivery time frame.

Therefore, Professor Tan suggests that expectations for moral education should serve as a **motivating force** rather than **a benchmark for evaluating real-world implementation**.

Efforts continue, both within China and abroad, to refine and enhance moral education models. In China, several systemic issues persist:

On the one hand, moral education remains overly singular. This singularity is evident in the source of instruction, primarily schools, which often neglect families and society, and in teaching methods that rely heavily on didactic, knowledge-transmission approaches. Although improvements have been made in recent years (e.g. through digital platforms that allow for interactive family involvement), challenges remain. The digital age introduces new risks such as exposure to harmful ideologies and problematic online behaviours, which must be addressed alongside technological integration.

On the other hand, Moral education suffers from low timeliness and weak outcomes. Currently, students' active engagement and intrinsic motivation are underdeveloped, which undermines the intended goals of moral education and may lead to adverse effects. Despite this, moral education remains a critical area of focus for primary and secondary education. Addressing practical issues, diversifying the types of providers, enhancing their quality, improving pedagogical strategies, and activating student agency are all essential to realising the genuine impact of moral education.

4. Educational tourisms and the New Moral Education Policy

Moral education requires integrating theory and practice and calls for more engaging and varied instructional methods to enhance student motivation and agency. In recent years, educational tourisms—a compulsory extracurricular practical subject for primary and secondary students—have gained significant recognition and support from government departments, schools, parents, and society.

Unquestionably, educational tourisms have become an essential channel for delivering moral education in China.

New moral education and educational tourism education goals are closely aligned from an educational objective standpoint. Both aim to cultivate students who can think independently, act effectively, reflect critically, and engage with society meaningfully.

From a content perspective, educational tourisms represent extracurricular, practice-based learning. **Components of moral education** serve as one of its primary implementation channels.

From a pedagogical and strategic perspective, educational tourisms tackle many challenges contemporary moral education faces, including their tendency toward uniformity, didacticism, and limited practical impact. Educational tourisms offer an innovative experiential model that parents and students accept more readily. They are characterised by high interactivity, learner-centred design, experiential learning, and an emphasis on stimulating thought—features that position them as a viable and promising pathway for enhancing the current state of moral education among primary and secondary school students.

2. **The "Double Reduction" Policy**
   a. **Opportunities Brought by the "Double Reduction" Policy**
   In July 2021, the General Office of the CPC Central Committee and the General Office of the State Council jointly issued the *Opinions on Further Reducing the Burden of Homework and Off-Campus Tutoring for Students in Compulsory Education*, commonly referred to as the **"Double Reduction" policy (双减政策)**. This policy explicitly limits the amount of homework assigned to students and mandates that all existing academic tutoring institutions be converted into non-profit entities. Furthermore, such institutions are prohibited from conducting subject-based training on national public holidays, weekends, and school vacations. As a result, primary and secondary school students are no longer subjected to weekend and holiday academic cramming.

   The announcement of this policy had an immediate and dramatic impact on China's private tutoring industry, leading to widespread layoffs, plunging stock prices, and in some cases, complete business closures.

   In sharp contrast, the **educational tourism industry** has experienced a different trajectory. With tutoring no longer dominating students' weekends and school breaks, there is a growing need for **knowledge-enriching and horizon-expanding alternatives**. Educational tourisms have encountered new opportunities and challenges under the "Double Reduction" framework as a hybrid model combining tourism and education.

Dr Han Yuanjun of the China Tourism Academy noted that the policy's implementation is likely to spur the rapid growth of the educational tourism market and transform it into a sector of significant economic scale.

According to Ctrip's *2021 Summer Vacation Tourism Big Data Report*, during the first summer following the release of the "Double Reduction" policy, educational tourism experiences emerged as a leading choice in customised family travel. Participation in educational tourisms during the summer holiday increased by over **650%** compared to the previous year, and educational tourism-related products accounted for nearly **70%** of all parent-child trip bookings. Similarly, data from the Lvmama platform showed that during the **2021 National Day holiday**, the number of educational tourism travellers increased by **78%** compared to the same period in 2020.

b. **Challenges Brought by the "Double Reduction" Policy**
While the **market potential** for educational tourisms is widely recognised, the current stage also reveals a growing challenge: many former tutoring institutions are **shifting their business focus** to the educational tourism sector. This influx is likely to intensify market competition significantly.

In response, educational tourism providers must adapt, upgrade, and innovate to **meet the evolving demands** of consumers and **secure their place** in a highly competitive industry landscape.

2.3. Educational tourisms and the socio-cultural, scientific, and ecological environments

Educational tourisms integrate research-based learning with travel experiences, capitalising on the vivid expressiveness, unique environmental features, and regional authenticity of a nation's social, historical, cultural, and natural geography. These programmes provide

immersive, multidimensional, experiential learning opportunities that engage the body and mind, nurturing moral character and personal growth.

Educational tourisms, characterised by exploration, practical engagement, interactivity, and research orientation, are particularly effective in helping primary and secondary school students develop independence, practical skills, management capabilities, and an innovative thinking mindset.

In the Chinese context, primary and secondary school educational tourisms often utilise resources such as traditional cultural heritage, including cultural relics, historical museums, and intangible cultural heritage, as well as modern science and technology infrastructures like science centres, laboratories, research institutes, high-tech enterprises, national parks, and nature reserves. These settings form the basis for designing educational tourism activities that are developmentally appropriate and educationally enriching for young learners.

Integration with Chinese Traditional Culture

Educational tourism programmes incorporating traditional Chinese culture have become a preferred choice for many schools. These programmes reflect shared goals and values, with mutually reinforcing content and methodologies. Integrating culture and practice in educational tourism curricula creates a unique synergy, enabling cultural education through hands-on experience.

By embedding elements such as core socialist values, red (revolutionary) culture, and intangible cultural heritage into course design, these tours provide students with opportunities to engage with Chinese culture experientially and reflectively. This fosters a deeper understanding of cultural essence and subtly enhances students' cultural confidence through firsthand encounters with heritage.

Integration with Scientific and Technological Development

Scientific and technological advancement is a key determinant of a nation's future. Science-themed educational tourisms have garnered strong support from both the government and society. These programmes enable students to engage in cutting-edge scientific experiences, sparking curiosity and encouraging the pursuit of truth. In doing so, students enhance their creative capacities and develop aspirations to contribute to national development through science and innovation.

Integration with Natural and Ecological Environments

Nature-based educational tourisms are increasingly recognised for their therapeutic and educational value. National parks and nature reserves strive to create high-quality curricula for nature educational tourisms to address the developmental needs of students of various ages. For example, in recent years, Chebaling National Nature Reserve in Guangdong Province has integrated science outreach and environmental education into its educational tourism course design. After thorough research and consultation, the reserve's science education team developed a curriculum that combines natural science education with ecological awareness.

Through a dual approach of "inviting students in" (to the nature reserve) and "reaching out" (to schools), the reserve allows more students to experience nature directly. These efforts promote respect for nature, nurture a love of the natural world, and strengthen commitment to environmental protection.

According to reserve managers, future initiatives will include developing more curricula by teams of science educators, volunteers, and interpreters and exploring "cloud classrooms" for nature education, thereby extending the benefits of nature-based learning to more young people and members of the public.

3. Microenvironmental Analysis of the Educational Tourism Market

3.1. *Companies, suppliers, and marketing intermediaries*

In the context of educational tourisms, suppliers refer to service providers. According to the important guideline document from the Ministry of Culture and Tourism, Service Specification for Educational tourisms (LB/T 054–2016), a supplier is "an entity that enters into a contract with the educational tourism organiser and provides services such as destination management, transportation, accommodation, catering, etc." Currently, China's educational tourism sector suppliers extend beyond conventional travel agencies. In addition to newly established educational tourism institutions, many traditional service providers have transitioned into the field, including off-campus tutoring institutions, tourist attractions, educational tourism bases (or camps), high-tech enterprises, and OTA (Online Travel Agency) platforms. Despite the impact of the

COVID-19 pandemic, by the end of 2021, tens of thousands of entities and companies still offered educational tourism services, indicating a massive supply scale.

It is common for educational tourism companies, suppliers, and marketing intermediaries to be part of the same corporate group. Supply channels encompass catering, accommodation, and transportation provided through destinations or tour bases. In contrast, marketing channels refer to the platforms used for promotion and sales, often involving third parties or intermediaries, such as OTA e-commerce platforms. At present, these actors remain in the **early stages of development**. Their collaboration is still exploratory and requires time to mature into a stable and efficient ecosystem.

3.2. *Educational tourism consumer segments*

a. **Primary Educational Tourism Groups**

The educational tourism market has a unique profile, with its core consumers being students, often organised in group formats. The majority of participants come from primary and secondary schools, as well as higher education institutions.

Comprehensive practical activities are mandatory in China's compulsory high school education curriculum guidelines. These activities are the same as academic subjects and crucial to the basic education curriculum system. Educational tourism programmes must be designed and marketed in alignment with the teaching goals of each school stage to meet students' actual developmental needs, from Year 1 of primary school to Year 3 of senior high school.

For university students who can choose their educational tourism experiences, programme design should focus on appealing to their interests and aspirations, thereby enhancing their motivation to participate.

b. **Business Educational Tourism Groups**

Business-oriented educational tourisms refer to learning programmes that involve visits to innovation-driven enterprises, focusing on language acquisition, professional skills training, and business communication. These programmes are typically aimed at adult learners, with learning objectives oriented toward career development and business engagement.

c. **Interest-Based Educational Tourism Groups**
Certain groups united by shared interests also create demand for educational tourisms. For example, artists interested in specific cultural themes may wish to visit renowned museums or galleries. At the same time, industry researchers may seek access to research institutions or conduct on-site investigations as part of their field studies.

3.3. *Competitors in the educational tourism industry*

Since implementing the "Double Reduction" policy, numerous private tutoring institutions have transitioned to the educational tourism sector, significantly boosting market competition. To ensure sustained and stable growth, educational tourism companies must utilise effective marketing strategies, often informed by successful case studies in the field.

a. **Meeting Consumer Needs Through Differentiated Educational Tourism Products**
Due to the emergence of new competitors, many small- and medium-sized educational tourism enterprises that lack adaptability or sufficient financial resources have already been displaced. Confronted with significant opportunities and considerable pressures, the educational tourism industry is undergoing a necessary transformation and upgrade process to adapt to the evolving commercial landscape. Over the next few years, the industry value chain is expected to be further optimised and refined. Driven by diverse consumer demands and the pursuit of higher quality, the market is moving toward greater segmentation, featuring finer divisions of labour and increasingly high-quality services tailored to public needs. As a result, educational tourism products are becoming more diverse, standardised, innovative, and locally contextualised. For example, Danxia Mountain Scenic Area in Shaoguan, Guangdong, has leveraged its resources to design various educational tourism products targeting different consumer groups. These include professional internships, educational excursions, and scientific research fieldwork.

b. Unified Planning and the Development of Educational Tourism Branding

Educational tourisms represent a fusion of tourism and education, and only when service quality and reputation reach a certain standard can a recognisable educational tourism brand be established. In recent years, educational tourism providers' previously unregulated and spontaneous growth has increasingly failed to meet consumer expectations. Simultaneously, public scrutiny of low-quality or misleading educational tourism products has intensified, prompting a shift in industry focus. Major organisations, powerful corporate groups, and local governments have begun to prioritise the construction of coherent and reputable educational tourism brands. In particular, some local governments have adopted a county-wide unified approach to brand development to build influence and increase competitive advantage.

A representative example can be found in Qufu, Shandong Province—the birthplace of Confucian culture. Qufu has effectively drawn upon its rich cultural heritage to develop a high-quality and well-regarded educational tourism brand. The Qufu Confucius Educational tourism Festival, a flagship cultural product promoted by the province, has become a central initiative for advancing educational tourism development. By integrating traditional Confucian culture with hands-on educational tourism experiences, the festival reinforces cultural education through immersive activities. As a result, Qufu now offers a range of high-quality and diversified educational tourism programmes, such as initiation and coming-of-age ceremonies, blessing rituals at the Confucius Temple, cultural campaigns linking memorisation of The Analects with free entry to heritage sites, pilgrimages, Yan Family cultural experiences, architectural exploration of the "Three Confucian Sites", and archaeological tours of the ancient Lu State capital (Li Cuifang & Jiang Aihua, 2018).

c. Integrating Social Resources to Achieve Win-Win Collaboration

The educational tourism ecosystem involves various stakeholders, including government agencies, schools, travel agencies, educational tourism bases (or camps), and educational training institutions. Given this complexity and diversity, educational tourism enterprises must enhance their communication and collaboration with these stakeholders. By integrating resources across sectors, companies can work towards establishing mutually beneficial business models that support sustainable development and long-term partnerships.

d. **Balancing Product Quality and Price to Achieve Competitive Pricing**
In the 21st century, consumers selecting educational tourism providers increasingly prioritise **value** for money over the lowest price. Many companies, camps, and tourist sites have devised creative pricing strategies in response to the vast and competitive educational tourism market. These include group discounts, free follow-up sessions, or waived entrance fees tailored to educational tourism participants. However, enterprises must maintain high service standards while offering these promotional prices. Ultimately, educational tourism providers can only build genuine price competitiveness and long-term consumer trust by balancing product quality with affordability.

e. **Diversified Marketing Channels and Targeted Promotion**
Today, most educational tourism companies adopt marketing strategies that combine digital channels with traditional media. Online platforms such as Weibo, WeChat public accounts, Douyin (TikTok), Xiaohongshu (Little Red Book), and video accounts are widely utilised for publicity and outreach. Many businesses also organise targeted live-streaming campaigns to engage specific market segments.

Well-executed and precisely targeted marketing initiatives can significantly enhance the visibility of educational tourism brands, expand a company's influence, and help establish a positive and recognisable brand image.

3.4. *The general public*

All corporate marketing activities occur within a community context, and businesses inevitably develop intricate relationships with the communities and publics they serve. Educational tourism enterprises are no exception.

In marketing their services, these companies must strive to align their corporate culture with the prevailing values of the community. Only by fostering this cultural alignment can they gain the trust and support of the public, ultimately securing a favourable reputation and a supportive environment for sustainable development (Guo Fangfang, 1990).

a. **Maintaining Positive Public Relations**
   Educational tourism enterprises must uphold an educational mission and create social value by addressing learners' practical needs as well as their intellectual and personal aspirations. To achieve this, educational tourism providers should establish and maintain strong public relations with stakeholders across various sectors of society. By promoting awareness of their cultural mission and societal contributions, companies can gain financial support, project opportunities, and positive public recognition. This, in turn, enhances their chances for collaboration, exposure, and capacity building, thereby strengthening their overall competitiveness.
b. **Maintaining Credibility and Public Trust**
   During marketing activities, educational tourism enterprises operate within two key environments: the task environment, which directly influences production and marketing, and the social environment, which, although not directly linked to daily operations, still plays a vital role in the company's survival and public standing.

   In today's media-driven landscape, a single piece of negative news can damage a company's credibility overnight. Students are the primary consumer group for educational tourisms, so the public holds exceptionally high expectations for quality, safety, and ethics. Therefore, marketing practices must remain truthful and grounded in reality. Exaggerated claims or misleading promotions may lead to a loss of credibility, and worse, may result in public complaints, legal exposure, or reputational damage.
c. **Maintaining Strong Brand Recognition and Reputation**
   A strong corporate image is a key indicator of successful marketing in the educational tourism industry. However, building a positive and trusted brand identity is not simple. As service-oriented providers, educational tourism companies must

consistently enhance customer satisfaction and promote positive word-of-mouth to raise both their brand awareness and brand favourability. To achieve this, companies must ensure that their educational tourism offerings demonstrate uniqueness, originality, and high quality. They can only develop lasting public goodwill and a recognisable, reputable brand within the community by maintaining these core attributes.

4. Chapter Summary

This chapter analyses the educational tourism marketing environment from both macro and micro perspectives. The macroenvironmental analysis includes factors such as domestic economic development, changes in household consumption patterns, the political landscape of the educational tourism industry, as well as social and cultural conditions, scientific and technological advancements, and the natural ecological context. The microenvironmental analysis focuses on educational tourism suppliers, marketing intermediaries, consumer groups, competitors, and the general public.

5. Post-lesson Exercises

1. Briefly describe the relationship between the **new moral education policy** and **educational tourisms**.
2. Under the **"Double Reduction" policy**, what **opportunities and challenges** have the educational tourism industry encountered?
3. In your opinion, what is the **future outlook** for the educational tourism market?
4. How can educational tourism enterprises **identify market positioning and breakthroughs** in the face of intense competition?

Chapter 4

Educational Tourism Market Segmentation and Consumer Analysis

Learning Objectives

1. To understand the **characteristics of different stages** of consumer groups in the educational tourism market.
2. To grasp the **methods used for analysing educational tourism market segments.**
3. To identify the **essential attributes of various educational tourism components** and understand the **objective patterns that govern their interrelationships**.

Learning Focus

1. Understand the concept of educational tourism market segments and the characteristics of different age-based customer groups.
2. Identify the purchase motivations of educational tourism consumers and the factors that influence these motivations.
3. Understand the decision-making process of educational tourism consumers.

Contents

1. Case Introduction

On the morning of 4 August 2022, several teachers, student representatives, and parent committee members from Fuxing Primary School in Guilin participated in a red-themed educational tourism organised at the Party-Masses Service Centre of the Guilin Liangxin Party Organisation.

Under the theme "Listening to Red Stories—Continuing the Red Bloodline and Inheriting Red Genes", the event was part of a summer volunteer service initiative for children. It was jointly organised by the Party-Masses Service Centre, the Women's Federation of Lingui District, the Guilin Liangxin Red Education College, the Retired

Military Affairs Bureau of Lingui, and other organisations. Hosted by the Women's Federation of the Tianshu Data Service Centre, the event was presided over by Vice Chairwoman Hai Ying from the district's Women's Federation.

During the event, teachers, students, and parents listened attentively to red stories and visited the site to better understand the historical significance of the Party-Masses Service Centre. Through this experience, students gained more profound insights into the spirit of service, strengthened their awareness of the Communist Party, and developed a sense of patriotism and civic responsibility.

The event also invited retired veterans to share their personal stories. One veteran vividly recounted their wartime experiences, helping students reflect on the hardships of past revolutionary times. Through emotional storytelling, students were inspired to strengthen their ideals and convictions, appreciate the peace and prosperity of contemporary China, and deepen their connection with red culture.[1]

1.1. *Reflection questions*

1. Based on the case above, what customer groups participated in the red-themed educational tourism? What are the characteristics of these groups?
2. What factors influenced the participation of Fuxing Primary School teachers and students in this red-themed educational tourism?

Educational tourism is an outdoor educational activity that combines comprehensive, hands-on, exploratory, and open-ended experiences. In a narrow sense, it refers to school-organised and education department-approved extracurricular activities that integrate research-based learning with travel-based experiences. These tours usually involve collective travel and accommodation, allowing students to participate in academic exploration and experiential activities. The primary participants are typically primary school students from across the country.

---

[1] *Source*: Relevant online media.

2. Concept and Characteristics of Educational Tourism Market Segments

In recent years, with the continuous advancement of quality-oriented education, schools have increasingly prioritised the development of educational tourism and experiential learning programmes. In many countries, educational tourism has become a regular component of primary and secondary school education, serving as an essential approach to support students' holistic development in mind and body. In China, relevant policies and measures have been actively introduced to ensure that educational tourism becomes an integral part of the national school curriculum, gradually evolving into a compulsory subject within primary and secondary education.

2.1. *Concept of educational tourism market segments*

Educational tourism is organised by schools based on regional characteristics, student age groups, and the requirements of specific subject areas. These tours involve students leaving the campus to travel in groups with shared accommodations, allowing them to experience life beyond the classroom. Through these experiences, students can broaden their horizons, enrich their knowledge, deepen their connection to nature and culture, and develop a sense of communal living and public morality.

Based on this definition, it is evident that the primary market segment for educational tourism consists of primary and secondary school students. However, their needs and desires vary by age. Therefore, effective educational tourism marketing must segment the market according to age, psychology, and academic stage, develop differentiated products, and adopt varied marketing strategies to cater to each educational level.

2.2. *The significance of analysing educational tourism market segments*

As an outdoor experiential educational activity, educational tourism is primarily designed for school-aged children, with their curriculum structured around natural resources, revolutionary (red) heritage, labour resources, national defence sites, and industrial facilities in society. The ultimate goal is to enhance students' comprehensive practical abilities. Generally, market segmentation refers to students' developmental traits, schools' educational philosophies, and the students' academic levels, ensuring that the tour's design aligns with its educational purpose. Specifically, analysing market segments in the educational tourism industry offers the following benefits:

1. **Helps identify target consumers and formulate marketing strategies:** By conducting segmentation analysis, educational tourism enterprises can identify suitable target consumers based on their business vision, access to educational tourism resources, and marketing capacity. This enables companies to further segment the market and develop customised marketing strategies tailored to the needs of specific consumer groups. Such analysis also provides more accessible and responsive data on customer needs. When consumption trends shift, businesses can swiftly adjust their strategies, recommend suitable destinations, or develop new educational tourism solutions to meet evolving demands. This enhances the company's agility and competitiveness in the market.

In 2020, the Central Committee of the Communist Party of China and the State Council issued the Opinions on Strengthening Labour Education in Primary, Secondary, and Tertiary Education in the New Era. The document emphasised integrating labour education across all educational levels, embedding it within family, school, and societal structures, and aligning it with moral, intellectual, physical, and aesthetic education. It further required that labour education be offered as a compulsory subject, with at least one period per week allocated in primary and secondary schools, and that schools establish guidelines for daily extracurricular and out-of-school labour activities.

In response, many educational tourism enterprises leveraged these policies and the schools' evolving needs to create labour-oriented curriculum frameworks. These frameworks included daily life skills, productive labour, and social service, integrating activities such as campus construction, base operations, and volunteer work. This approach allowed students to engage in labour education comprehensively while realising a collaborative school–family–society engagement model.

2. **Facilitates the discovery of market opportunities and expansion into new markets:** Conducting an educational tourism market segment analysis enables enterprises to better understand consumer

demographics, including regional distribution, demand preferences, purchasing potential, satisfaction levels, previous educational tourism experience, and travel habits. A comparative analysis of different market segments helps businesses identify untapped opportunities, detect emerging patterns in consumer needs, and respond proactively.

By seizing these opportunities and aligning them with the company's existing operational strengths, enterprises can develop new educational tourism products, formulate updated marketing plans, and create a reserve of offerings suited to future demand. This grants companies greater initiative and control over product renewal and upgrading, enabling them to stay competitive and responsive to evolving market trends.

3. **Enables the concentrated allocation of resources to target markets:** Every enterprise operates with limited human, material, and financial resources. After analysing market segments, selecting the most suitable target market enables a company to focus its resources strategically, gain a competitive edge in specific market niches, and make optimal use of available resources. Through this process, businesses can amplify their strengths, avoid vulnerabilities, and enhance their competitiveness, ultimately securing a stronger position within their chosen target market.

2.3. *Analysis of years 1–3 in primary school*

1. **Analysis of years 1–3 in primary school:** At this stage, students are primarily interest-driven learners who have not yet developed consistent study habits and tend to engage in passive learning. Therefore, educational tourism products designed for this age group should focus on stimulating students' learning interests and fostering habit formation. One practical approach is the five-sense teaching method, which engages students through multi-sensory experiences. For example, activities might include observing the natural colours of plants, listening to animal sounds, and creating nature journals. These practices help foster curiosity about the natural world while enhancing students' logical thinking and expressive abilities.
2. **Analysis of years 4–6 in primary school:** Having built upon earlier experiences, students at this stage typically develop stronger

observational, imaginative, cognitive, and linguistic abilities. Their curiosity and desire for knowledge intensify, leading them to demonstrate a capacity for independent thinking and exploratory behaviour. Educational tourism at this level should help students form reasonable interpretations of real-life phenomena and guide them in expressing their thoughts creatively. The aim is to encourage self-directed exploration and reflective engagement.

3. **Analysis of years 1–3 in lower secondary school (junior high):** Students experience notable changes in physical development, knowledge acquisition, and psychological maturity by the lower secondary stage. However, they still retain many characteristics of younger learners. Their thinking remains intuitive and emotionally driven, with a limited capacity for abstract reasoning. At the same time, their self-awareness increases, and they begin to seek recognition from peers and adults, developing aspirations for independence and achievement.

As students transition into secondary school, they encounter shifts in academic content, expectations, methods, and learning environments. Educational tourism programmes should, therefore, focus on extracting meaningful and personally relevant elements from the curriculum to foster students' intrinsic motivation. The goal is to help them understand, retain, and apply knowledge through experiential learning.

4. **Analysis of years 1–3 in upper secondary school (senior high):** The upper secondary years represent a critical period in adolescent development, during which students' moral character, worldview, self-identity, emotions, personality, and values are actively forming. Students at this stage often exhibit a blend of maturity and immaturity, independence and dependence, self-awareness and impulsivity, resulting in a complex psychological profile.

To meet their developmental needs, educational tourism activities should draw on the "symbiotic effect", leveraging the power of group dynamics. Activities should provide opportunities for students to take responsibility, discover personal strengths, and thrive within a collective setting. Programmes can be designed around confidence-building, resilience, overcoming setbacks, and embracing challenges to support their growth into motivated and self-assured individuals.

2.4. *Characteristics of educational tourism market segments*

1. **Analysis of years 1–3 in primary school:** At this stage, students are primarily interest-driven learners who have not yet developed consistent study habits and often engage in passive learning. Therefore, educational tourism products designed for this age group should focus on stimulating students' learning interests and fostering the formation of study habits. One practical approach is the five-sense teaching method, which involves engaging students through multi-sensory experiences. For example, activities might include observing the colours of plants in nature, listening to the sounds of animals, and creating nature journals. These practices help cultivate curiosity about the natural world while enhancing students' logical thinking and expressive abilities.
2. **Analysis of years 4–6 in primary school:** Building on earlier experiences, students at this stage typically develop enhanced observational, imaginative, cognitive, and linguistic abilities. Their curiosity and thirst for knowledge intensify, and they exhibit a capacity for independent thinking and exploratory behaviour. Educational tourism at this level should assist students in forming sound interpretations of real-life phenomena and guide them in expressing their thoughts creatively. The aim is to promote self-directed exploration and reflective engagement.
3. **Analysis of years 1–3 in lower secondary school (junior high):** Students experience notable changes in physical development, knowledge acquisition, and psychological maturity by the lower secondary stage. However, they still retain many characteristics of younger learners. Their thinking remains intuitive and emotionally driven, with limited capacity for abstract reasoning. At the same time, their self-awareness increases, and they begin to seek recognition from peers and adults, developing aspirations for independence and achievement.

As students transition into secondary school, they encounter shifts in academic content, expectations, methods, and learning environments. Educational tourism programmes should therefore focus on extracting meaningful and personally relevant elements from the curriculum to foster students' intrinsic motivation. The goal is to help them understand, retain, and apply knowledge through experiential learning.

4. **Analysis of years 1–3 in upper secondary school (senior high):** The upper secondary years are a critical period in adolescent development, during which students' moral character, worldview, self-identity, emotions, personality, and values are actively forming. Students at this stage often display a mix of maturity and immaturity, independence and dependence, self-awareness and impulsivity, resulting in a complex psychological profile.

To meet their developmental needs, educational tourism activities should draw on the "symbiotic effect", leveraging the power of group dynamics. Activities should provide opportunities for students to take responsibility, discover personal strengths, and thrive within a collective setting. Programmes can be designed around confidence-building, resilience, overcoming setbacks, and embracing challenge to support their growth into motivated and self-assured individuals.

2.5. *Steps in analysing educational tourism market segments*

In the process of identifying and evaluating market opportunities and selecting target markets, educational tourism marketing managers must not only thoroughly analyse the broader marketing environment and develop a general understanding of consumer markets, producer markets, intermediary markets, and government markets, but also conduct marketing research, gather information, and perform market measurements and forecasting. These efforts guide decisions on which new educational tourism products should be developed and help identify consumer demand within target groups.

The analysis of educational tourism market segments typically involves the following steps:

1. **Analyse the characteristics of educational tourism consumers:** Educational tourism business operators must establish consumer profiles for their primary product candidates, regardless of whether those candidates are individuals, groups, or existing or potential customers. This involves answering several key questions: Who buys educational tourism products? Who uses them? How are they used? Where can they be purchased? How are they bought? Why are they purchased? When are they bought?

By addressing these questions, businesses begin to identify their **target consumer groups**. New entrants to the market typically rely on their own marketing experience, while providers entering schools should prioritise **on-site market research**. They should engage directly with schools to understand their educational values, teaching priorities, previous educational tourism experience, and student development needs. These insights help determine consumer expectations and guide the selection of target markets.

2. **Identify the basic needs of potential consumers:** Understanding the **basic needs of potential consumers** is crucial. Market research can uncover the core expectations of prospective customers. For example, a school in Hunan Province wanted to organise a **two-day educational tourism** using local resources to ignite Year 7 students' interest in geography. After identifying this demand, the educational tourism provider selected suitable destinations, tailored the itinerary accordingly, and implemented marketing strategies to align with these needs, increasing the likelihood of recognition and purchase.
3. **Understand the needs of different consumer groups:** Educational tourism consumers exhibit a wide range of demands. Even for the same product type, different groups may emphasise different needs. For example, under labour education policies, all students must complete **at least one period of labour practice per week**, but their focus varies by academic stage:
   - Lower primary students should be introduced to labour awareness, basic self-care, and the joy of working.
   - Upper primary students should cultivate hygiene habits, assist with household chores, and engage in school and community service.
   - Secondary students should acquire practical labour skills, study home economics, participate in community service, and develop responsibility and vocational awareness.

Even with unified policy goals, students of different ages and schools have varied expectations for these labour-themed educational tourism courses.

4. **Use key differentiated demands as preliminary criteria:** When analysing market segments, companies may extract shared consumer

needs while focusing particularly on differentiated needs. These differentiating factors—such as geographical, psychological, behavioural, and school-specific elements- form the basis for market segmentation because they reflect variations in consumer demand. Following are some of the most common segmentation criteria:

i. **Geographical factors:** Segmenting by geography involves dividing the educational tourism market by country, region, province, city, or physical terrain such as mountains, plains, lakes, and deserts. Geographical features influence consumption preferences due to variations in climate, transportation, communication infrastructure, cultural traditions, and economic development. For instance, schools in different parts of China have diverse educational tourism needs based on regional resources. A school in Changsha, Hunan, may prefer tours that include visits to rural or ethnic minority communities to help students grasp urban–rural differences.
ii. **Psychological factors:** Consumers with similar demographics may respond differently to the same product due to psychological differences. These include the following:
    - *Lifestyle*: Daily, work, and entertainment consumption habits influence product needs and usage behaviours. Even for the same product, preferences may vary in design, size, or appearance.
    - *Social Class*: People from various social classes originate from different backgrounds and interests, which influence their product needs.
    - *Personality*: Unique personality traits create consistent patterns in purchasing behaviour. Personality can be a strong foundation for segmentation when linked to brand preference.
    - *Preferences*: Consumers demonstrate differing levels of preference for various themes—e.g. military education, science exploration, nature appreciation, or cultural heritage. Some have strong preferences, some moderate, while others have none. To grow or maintain market share, businesses must understand these preferences and tailor their products and services accordingly.

5. **Preliminary selection of potential consumer groups:** The elimination method can be applied during the preliminary selection to progressively

narrow the pool of educational tourism consumers. Starting with the most evident criteria, businesses can exclude one group at a time based on specific factors. This process is repeated across additional criteria, continually refining the selection until only a few candidate groups remain—these form the initial target consumer groups.

Two major pitfalls must be avoided in this stage:

1. Valuable business opportunities are missed by overlooking market segments that offer substantial potential for the company's core products.
2. Spending excessive time and resources investigating markets that do not have future viability.

To ensure effectiveness, preliminary selection should be applied across all available markets. The process should also be conducted with speed and cost efficiency in mind. Companies should extract relevant data from widely available public sources whenever possible to streamline early-stage decision-making.

2.6. *Methods of analysing educational tourism market segments*

Methods for analysing the educational tourism market can generally be categorised into two:

1. **Qualitative analysis:** Qualitative analysis involves collecting data through historical reviews, literature analysis, interviews, observations, and first-hand participation. These methods rely on non-numerical techniques to interpret information and draw research conclusions. It is particularly useful for exploring consumer perceptions, preferences, behavioural motivations, and contextual factors influencing educational tourism demand.
2. **Quantitative analysis:** Quantitative analysis, in contrast, produces results expressed through numerical data. Research in this category is designed to enable researchers to systematically compare and analyse data to draw valid and generalisable conclusions.

In educational tourism market research, quantitative analysis primarily relies on questionnaire surveys. A questionnaire is a structured set of questions designed to gather specific information aligned with the research objectives. It is one of the most vital tools in market research. The quality of the questionnaire design directly impacts the accuracy and reliability of the collected data.

A complete questionnaire typically includes three sections:

i. *Introduction*: This section should be brief, persuasive, and captivating to quickly attract the respondent's interest. It typically includes the title, purpose, and significance of the survey, along with instructions for completion.
ii. *Main Body*: This section forms the core of the market research and includes key questions designed to gather essential information. These questions must be carefully crafted and presented with appropriate answer options.
iii. *Survey Record*: This refers to documenting relevant details during the survey process. It typically includes the name or ID of the surveyor, the date and location of the survey, the name, address, and contact number of the respondent, as well as the name of the verifier and the questionnaire ID.

The questionnaire design process generally involves eight standard steps, as listed in Figure 1.

1. **Clarify the purpose and content of the survey**
2. **Determine methods for data collection**
3. **Identify the types of questions**
4. **Determine strategies for addressing the questions**
5. **Establish the sequence of questions**
6. **Evaluate the questionnaire**
7. **Conduct a pilot survey and make revisions**
8. **Develop the final version of the questionnaire**

Figure 1. Eight steps of questionnaire design.

Once the questionnaire survey is completed, the collected data must undergo quantitative analysis. Common quantitative methods include frequency distribution, hypothesis testing, cross-tabulation, analysis of variance (ANOVA), correlation analysis, and regression analysis.

(1) **Frequency distribution:** Frequency distribution is a mathematical method used to record how many times a variable assumes each of its values, typically presented as percentages. The frequency distribution of a variable is essentially a table displaying the count, rate, and cumulative percentage of each value.

This method effectively clarifies the relationship between variables and reveals data characteristics. However, due to the sometimes-overwhelming amount of information, researchers often rely on descriptive statistics to summarise data. Standard descriptive measures include the following:

- *Measures of central tendency*: mean, mode, median;
- *Measures of dispersion*: range, interquartile range, variance, standard deviation.

(2) **Hypothesis testing:** Hypothesis testing can be categorised into parametric and non-parametric tests based on the type of variables:

- Parametric tests are applied to interval and ratio variables and typically include *t*-tests and *z*-tests, which assess hypotheses concerning one or two population means.
- Nonparametric tests are used for nominal and ordinal variables.

(3) **Cross-tabulation:** Cross-tabulation is a statistical technique that presents two or more variables simultaneously in a table format, illustrating their joint distribution. While frequency distributions describe a single variable at a time, cross-tabulation enables the combined analysis of multiple variables. It is widely employed in market research due to its advantages:

- offers clear and intuitive outcomes for decision-makers;
- illustrates the connection between findings and business decisions;
- aids in explaining complex issues;

- addresses missing data issues;
- simplifies complex datasets.

The quality of the crosstab relies on the researcher's skill in selecting key variables and designing the tables accordingly. These variables should be defined before the commencement of data collection.

However, cross-tabulation also has two limitations:

1. If multiple variables are taken into account, a large sample size is necessary.
2. It is challenging to ensure that all relevant variables are analysed, which may lead to incorrect conclusions or overlooked relationships, even if the correct variables are used.

(4) **Analysis of variance (ANOVA):** ANOVA is used to determine if two or more sample means differ significantly. It includes the following:

- one-way ANOVA (single factor);
- two-way or multi-factor ANOVA (multiple factors).

(5) **Correlation analysis:** In market research, researchers are often interested in whether two interval or ratio variables are related and how strong that relationship is. For example, they might examine the following:

- the relationship between market share and advertising expenditure;
- the link between brand attitude and price sensitivity.

This type of analysis is known as correlation analysis, and the strength of the relationship is measured using the correlation coefficient. The coefficient ranges between −1 and 1:

- A value close to 1 indicates a strong positive correlation.
- A value close to −1 indicates a strong negative correlation.
- A value of 0 indicates no correlation.

For example, suppose a study finds a correlation coefficient of −0.8776 between consumers' age and product rating. This suggests a strong negative correlation: younger consumers tend to give higher ratings, indicating that the product is more popular among younger age groups.

(6) **Regression analysis:** Regression analysis is commonly used to explain and predict factors such as the following:

- market share;
- sales performance;
- brand preference;
- marketing effectiveness.

A mathematical function expresses the quantitative relationship between one dependent variable and one or more independent variables. There are two main types:

i. simple linear regression (one independent variable);
ii. multiple linear regression (two or more independent variables).

3. Educational Tourism Consumer Motivation

Educational tourism enables students to engage with diverse educational approaches beyond traditional knowledge-based instruction. These experiences focus on practical engagement—learning through doing and doing through learning—thereby enhancing understanding and transforming theoretical, "static" knowledge into "living" practice.

For example, when secondary school geography textbooks introduce the cultural diversity of China, including the Han, Miao, Manchu, Hui, and Tujia ethnic groups, the corresponding educational tourism may take students into ethnic minority villages. This experience allows them to gain firsthand insights into traditional clothing, food, customs, festivals, and arts, thereby deepening their understanding through environmental immersion. Similarly, content on bronze artefacts and oracle bone script in history textbooks may remain abstract without visual engagement. However, suppose students visit a museum to examine the patterns, materials, and inscriptions on ancient bronzeware, such as the Four-Rams Vessel or the Simuwu Ding, and participate in related interactive activities. In that case, their learning becomes multi-dimensional and significantly enriched.

3.1. *The concept and development of educational tourism*

The term "educational tourism" first appeared in official Chinese policy in the Outline for National Tourism and Leisure (2013–2020), issued by the General Office of the State Council in 2013. Educational tourism emerged from China's longstanding **educational tourism traditions**, incorporating modern pedagogical objectives. In 2014, Wang Dinghua, Director of the Department of Basic Education at the Ministry of Education, defined the concept of "educational tourism" and proposed four core characteristics:

1. encouraging direct student participation and engagement;
2. group activity formats;
3. non-spontaneous (i.e. formally arranged);
4. distinct from extracurricular interest clubs or school competitions.

The Opinions on Promoting Educational Tourism for Primary and Secondary School Students (2016) outlined how government departments and schools should organise educational tourism collaboratively. These tours will be conducted off-campus, involving shared accommodation and research-based learning, representing a blended approach to on- and off-campus education and an effective method for comprehensive practical learning. Educational tourism thus integrates education and leisure, creating a new tourism experience.

3.2. *Educational tourism and consumer motivation*

1. **The concept of motivation:** Motivation is the internal psychological process or drive that stimulates and directs individuals toward a specific goal. It arises from needs that act as latent driving forces and manifests in forms such as intentions and desires. Motivation is activated when a suitable incentive appears, transforming a latent need into action.

While needs and motivation are closely related, they are not identical. A vague, undifferentiated need is referred to as an intention. At this stage, one may know a direction for action, but the specific need or how to satisfy it is unclear. A desire, on the other hand, is a consciously acknowledged and wanted need. However, it does not qualify as motivation unless the desire results in actual behaviour. The need becomes a motivational force only when it drives and sustains action.

First introduced in psychology in 1918, the concept of motivation serves three primary functions:

i. **Activation:** initiating behaviour;
ii. **Direction:** guiding behaviour toward a goal;
iii. **Regulation and maintenance:** sustaining the behaviour over time.

Motivations can be classified by origin as

- physiological motivation;
- social motivation;

or by cause as

- **intrinsic motivation** (e.g. ideals and aspirations);
- **extrinsic motivation** (e.g. rewards and punishments);
- **participation motivation** (defined as the psychological tendency or drive to engage in an activity, which arises from internal or external factors).

2. **Conditions for motivation to arise:** The formation of motivation requires the joint influence of internal and external factors:
    - The internal condition refers to an unfulfilled need that drives the individual to pursue satisfaction, creating motivation. Needs serve as the source of behavioural initiative, while motivation represents their specific expression.
    - The external condition serves as a stimulus or incentive capable of satisfying the need. The stronger the intensity of the stimulus, the more likely it is to evoke powerful motivation and trigger corresponding behaviour.

3. **Main motivations of educational tourism consumers:** With the advancement of general tourism motivation theory, researchers have begun to explore the specific motivations of educational tourism participants. For instance, Liang Jian investigated international students

in Chongqing and found that cultural education was the primary push factor, while the region's educational advantages were significant pull factors. Personal preferences and regional development also influenced behaviour. Dong Jianying *et al.* surveyed students in Taiyuan and identified motivations, including knowledge-seeking, physical and emotional well-being, self-fulfilment, and attraction to specific educational tourism experiences. Li Donghe *et al.* conducted questionnaires and interviews in three representative primary and secondary schools in Hefei, identifying three dimensions of motivation: perceived benefits, environmental awareness, and perceived barriers. Zhang Jiaojiao studied university students and categorised their motivations into five types: interpersonal (opposite-sex) socialising, relief from familiar environments, the challenge of independent living, exotic consumption, and avoidance of social interference.

Synthesising domestic and international research, this book identifies the following key motivations for educational tourism participation:

1. **Knowledge-seeking motivation:** Educational tourism provides an experiential, outdoor learning process that enables students to engage with society and expand their knowledge. Entering unfamiliar contexts cultivates a sense of novelty and growth. Educational tourism offerings include red (revolutionary) education, labour education, military simulation, natural science, cultural heritage, and volunteer service. Each experience broadens students' intellectual horizons and enriches their life experiences.
2. **Tourism motivation:** Tourism motivation emerges as a product of social and economic development, representing the individual's desire to travel for specific purposes. Dann,[2] an early tourism motivation theorist, proposed that both push and pull factors influence tourism:
    - *Push factors*: psychological motivations such as stress relief or socialisation;
    - *Pull factors*: destination-based attractions such as natural beauty and cultural uniqueness.

---

[2]Dann, G.M.S. (1977). Anomie, ego-enhancement and tourism. *Annals of Tourism Research*, 4(4), pp. 184–194.

When educational tourism activities are framed as mandatory educational activities, societal encouragement enhances students' desire to travel. Furthermore, ethnic diversity, local customs, and religious traditions can be powerful cultural motivators driving participation.

3. **Health motivation:** Health motivation is the drive to alleviate academic stress and enhance physical and emotional well-being. Classroom-based learning can lead to tension and fatigue due to its rigid structure. Educational tourism offers a change of environment, allowing students to relax, engage in hands-on activities, and boost their mental and physical health.
4. **Practical development motivation:** This includes the drive to broaden perspectives, cultivate emotional maturity, and develop character. Educational tourism is particularly effective for expanding knowledge, fostering innovation, and nurturing curiosity, a high-level need for many students. Activities such as hiking, nature education, military simulations, tech innovation, and historical exploration can satisfy this need by providing challenges and novel experiences that allow students to discover new goals and understand diverse cultures.[3]

4. Factors Influencing the Purchase Motivation of Educational Tourism Consumers

The purchase motivation of educational tourism consumers refers to the subjective conditions and internal driving forces that lead to educational tourism participation. However, such behaviour is also dependent on objective conditions. Even when consumers intend to participate in educational tourism, when their physical and psychological conditions are suitable, and when the destination aligns with their interests and academic level, certain external conditions must still be met to activate and reinforce purchase motivation.

[3]Dong, J. and Ren, L. (2016). A study on the motivations of secondary school students for study tour participation based on principal component analysis: A case study of Taiyuan. *Economic Issues*, (7), pp. 119–124.

In the educational tourism market, the factors that most significantly influence consumer purchase motivation include the following eight factors:

1. **Cost of educational tourism products:** The cost of the educational tourism product serves as the foundation for pricing decisions and must be regarded as a primary consideration. The price of a product reflects its inherent value, which, in the context of educational tourism, is determined by both the worth of the purchased components and the quality and quantity of services provided by the organising institution or base (camp).

   The cost structure comprises fixed costs and variable costs:

- **Fixed costs** are unaffected by changes in the scale of operations over a specific period or scope. These include expenses such as property rental, depreciation of facilities, fixed salaries, marketing and promotion expenses, telecommunication costs, and other fixed-asset depreciation.
- **Variable costs** fluctuate with sales volume and include expenditures such as instructor fees, sales commissions, transportation, accommodation, meals, admission tickets, educational tourism equipment, and printed materials. Variable costs typically constitute a significant proportion of the total pricing structure.

2. **Business development strategies of educational tourism enterprises:** The strategic orientation of an educational tourism business directly influences pricing. There are generally three types of development strategies:
   - intensive development strategy;
   - integrated development strategy;
   - diversified development strategy.

Among these, the intensive development strategy significantly affects pricing. When an educational tourism enterprise aims to capture new market share or enter new target markets, it often needs to adjust or re-establish its pricing to align with changing market demands.

3. **Characteristics of the educational tourism product:** The features of the educational tourism product offered by various service providers vary significantly, influencing pricing decisions to different extents. The more complex and comprehensive the product or service bundle, the less control marketers may have over pricing fluctuations. Regarding educational tourism destinations, when a product is easily substitutable, its price elasticity increases. This raises the likelihood of price competition and discounting among providers.
4. **Market competition in the educational tourism sector:** Market competition within the educational tourism industry is most evident in price competition among peers. The term "peer pricing" refers to the pricing of similar educational tourism products by other institutions or camps within the same market. Peer pricing is referenced for two main reasons:
   1. Education authorities and schools often compare prices when organising student tours, developing a sense of standard pricing expectations.
   2. Companies may strategically set prices above or below market averages, depending on current market conditions and their development strategy.

For instance, if an educational tourism product is unique, service providers must compare it closely with competing products to identify advantages and disadvantages and evaluate market demand and influence to justify final pricing decisions.

5. **Influence of national policies:** As part of China's broader effort to implement moral education goals and promote holistic student development, educational tourism—an emerging model that fuses education with tourism—has received strong policy support from both the state and local governments.

Since 2013, various provinces and municipalities have conducted pilot programmes. In 2014, the State Council approved the incorporation of educational tourism into student quality education. From 2016 to 2017, the National Tourism Administration and the Ministry of Education introduced a series of policies, providing a solid platform for the growth of educational tourism and its integration into the national education system.

6. **Influence of the social environment:** Evidence shows that educational tourism relies heavily on social resources and the surrounding environment. The environment serves not only as a backdrop but also as an essential medium for the experience. At the same time, educational tourism has a significant impact on the environment. For example, the COVID-19 pandemic of 2020 severely affected the tourism industry, leading to a temporary standstill in the educational tourism sector. This highlights the sector's vulnerability to major social and public health disruptions, underlining the need for adaptability and contingency planning.
7. **Influence of safety factors:** As the educational tourism sector expands rapidly, safety has become the most critical factor affecting its sustainable development. In 2016, the Ministry of Education and 10 other departments issued the Opinions on Promoting Educational Tourism for Primary and Secondary Students, emphasising the safety principle first. The document stated: "Educational tourisms must adhere to the principle of safety, establishing comprehensive safety mechanisms, clarifying responsibilities, and implementing effective measures to ensure student safety". Local governments were instructed to develop scientific and practical safety assurance plans, establish effective mechanisms for responsibility allocation, incident handling, and dispute resolution, and enforce tiered registration systems to ensure full accountability at every level.

When consumers develop an intention to participate in educational tourism and meet the necessary conditions, they still face choices among various products, such as selecting the destination, the organiser, and the travel mode. If a destination is associated with potential risks such as political instability, economic disruption, or natural disasters, consumers will likely abandon it in favour of safer alternatives.

8. **Influence of weather conditions:** Weather is an unpredictable factor influencing consumer decisions in the educational tourism market. In extreme or adverse weather conditions, the number of participants tends to decrease substantially, and consumers may adjust their travel plans accordingly, including changing destinations, rescheduling activities, or cancelling participation altogether.

5. The Purchase Decision Process of Educational Tourism Consumers

In the field of consumer behaviour research, the information-processing model—also known as the rational decision-making model—has long held a dominant position. However, in the 1980s, scholars such as Holbrook began to advocate for a deeper focus on consumer emotion, arguing that consumer behaviour studies must also account for personal experience and the emotional aspects of decision-making. According to this view, the consumer is a "thinker" and a "feeler". In other words, when an educational tourism consumer decides to purchase an educational tourism product, they seek the educational value of the experience and the emotional and sensory gratification that the experience brings.[4] As a practical and experiential activity, the process by which consumers purchase educational tourism products or services typically involves five key stages, as illustrated in Figure 2.[5]

Across these five stages, need recognition is driven by the consumer's emotions; emotionally oriented solutions guide information search; evaluation and selection of educational tourism options are based on comparisons between emotional solutions; purchase plan formulation reflects emotional preferences; and post-purchase behaviour stems from the consumer's emotional evaluation of the experience. Every stage of the educational tourism is grounded in the consumer's decision-making process, which is grounded in emotion.

1. **Recognition of need:** Recognition of need arises when educational tourism consumers perceive an unmet learning requirement. This unmet need originates from the gap between a consumer's ideal and

Figure 2. The process of consumer purchase decision-making: Need recognition.

---

[4]Liao, Y. (2010). *A Study on the Purchase Decision Process and Influencing Factors of Experiential Consumption*. Wuhan: Wuhan University Press.

[5]Wang, S. and Zhang, J. (2016). *Consumer Behaviour: Analysis and Practice*. Beijing: China Renmin University Press.

actual states. The ideal state refers to the conditions or experiences the consumer wishes to attain, whereas the actual state reflects the consumer's current perception of their situation and feelings. Problem recognition occurs when the gap between the ideal and actual states becomes substantial enough to prompt a purchase decision, as shown in Figure 3.

2. **Information Search:** The decision-making process for purchasing an educational tourism product starts with the consumer recognising a specific need. This need may stem from an internal physiological drive, an external environmental stimulus, or a combination. Once consumers know their need for an educational tourism product, and if a product can satisfy this need, targeted purchase motivation will arise.

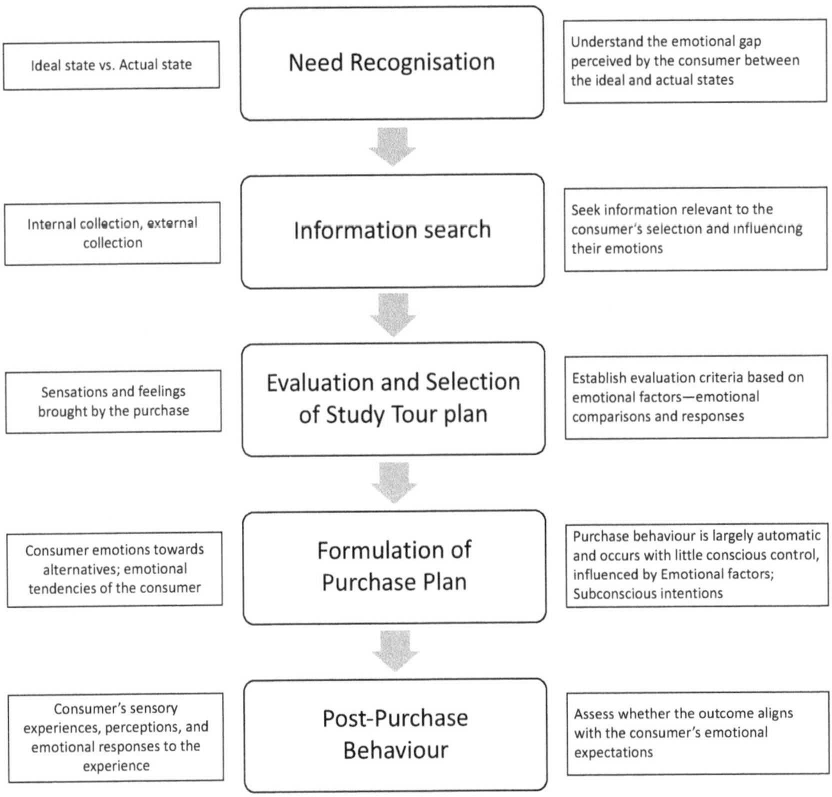

Figure 3. Model of the purchase decision process and its influencing factors.

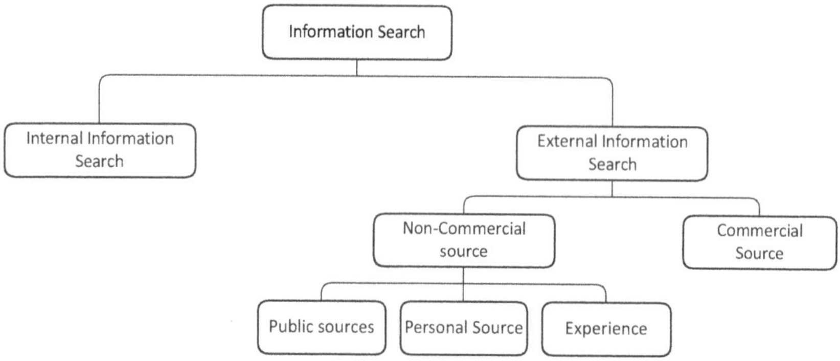

Figure 4. Consumers of educational tourism gather information.

Some of these needs can be met through purchasing behaviour, and the necessary information may already be known to the consumer from previous experiences or prior information searches. However, specific needs cannot be fulfilled or assessed solely based on experience, thus requiring additional information gathering.[6]

Consumers of educational tourism gather information from various sources, which are typically classified as internal and external. Internal sources primarily include the consumer's previous experiences and impressions of purchasing similar educational tourism products. External sources are generally divided into the following four categories (Figure 4):

i. *Commercial sources*: This includes advertising, sales personnel, distributors, educational tourism exhibitions, packaging, and promotional materials.
ii. *Public sources*: Including WeChat public accounts, video accounts, live-streaming platforms, and editorial commentary on educational tourism products.
iii. *Personal sources*: Such as family members, friends, neighbours, and colleagues.
iv. *Experiential sources*: Such as trial usage and direct experience.

[6]Dong, Q. and Zhang, R. (2018). *Tourism Marketing Practice*. Beijing: Beijing Institute of Technology Press.

3. **Evaluation and selection of educational tourism programmes:** After gathering information about educational tourism products from various channels, consumers typically organise and systematise this data before conducting a comparative analysis and evaluation. In the evaluation and selection process, consumers primarily consider factors such as the content of the educational tourism product, brand reputation and image, pricing, safety standards, expected learning outcomes for students, and the logistical support provided during the educational tourism.
4. **Formulating the purchase plan:** After evaluating and selecting educational tourism programmes, consumers intend to purchase a specific brand—that is, they will develop the willingness and plan to buy. However, having a purchase intention does not necessarily mean that consumers will go ahead and make the purchase. Three factors may influence a consumer's final purchasing decision: others' attitudes, unexpected circumstances, and perceived risks.[7]

Perceived risks that educational tourism consumers may face when purchasing products generally fall into the following categories:

- *Functional risk*: This refers to the possibility that the purchased educational tourism product or service may fail to deliver the promised effects or anticipated results. For instance, students may embark on a science-themed educational tourism, only to discover that the actual learning outcomes do not align with the promotional claims made by the organiser.
- *Financial risk*: This refers to the concern that consumers may invest substantial amounts in an educational tourism product yet fail to receive quality service or value in return, commonly referred to as "wasting money".
- *Safety risks*: Safety risk refers to potential safety issues from purchasing an educational tourism product. For instance, during a two-day educational tourism, students may fall into the water due to improper handling if they partake in water-based outdoor activities. Likewise, students may face safety incidents during overnight stays if they leave the premises without permission at night.

[7] *Ibid.*

- *Psychological risk*: This refers to the possibility that consumers may experience dissatisfaction or disappointment after purchasing a product or service. For example, suppose students join an educational tourism at a popular destination during the Labour Day "Golden Week" holiday. In that case, the high volume of tourists may lead to delays or discomfort in transportation, accommodation, and activities, ultimately diminishing the overall experience.
- *Post-purchase behaviour*: The purchase process does not end once an educational tourism consumer has bought the product; instead, it transitions into the post-purchase stage. During this phase, the consumer participates in educational tourism and assesses their level of satisfaction through direct experience. Additionally, after the educational tourism concludes, consumers often engage in activities such as evaluating learning outcomes, presenting findings, and showcasing achievements. At this point, the consumer is also faced with whether to use the same product or service again.

6. Chapter Summary

This chapter has outlined the key concepts related to educational tourism market segmentation, including methods of analysing target consumer groups and consumer purchasing behaviour. By applying the theoretical knowledge presented in this chapter to case study analysis, one can gain a deeper understanding of the characteristics of educational tourism consumers, enhance analytical skills, and lay a solid foundation for identifying and addressing the target market in the educational tourism industry.

7. Post-chapter Exercise

Using a specific region as an example, conduct a quantitative analysis of the educational tourism market consumer group using a questionnaire survey. Summarise the needs and characteristics of that region's educational tourism market consumer group.

Chapter 5

Marketing Channels and Expansion Strategies in the Educational Tourism Market

Learning Objectives

1. To comprehend the relationship between educational tourism products and the educational tourism market, and the significance of broadening marketing channels.
2. To gain insight into methods for expanding marketing channels for educational tourism products across different regions.
3. To master pricing strategies for educational tourism products and approaches to expanding marketing channels, both online and offline.

Learning Focus

1. The relationship between educational tourism products and the educational tourism market.
2. Pricing strategies for educational tourism products and methods for expanding marketing channels across different regions.

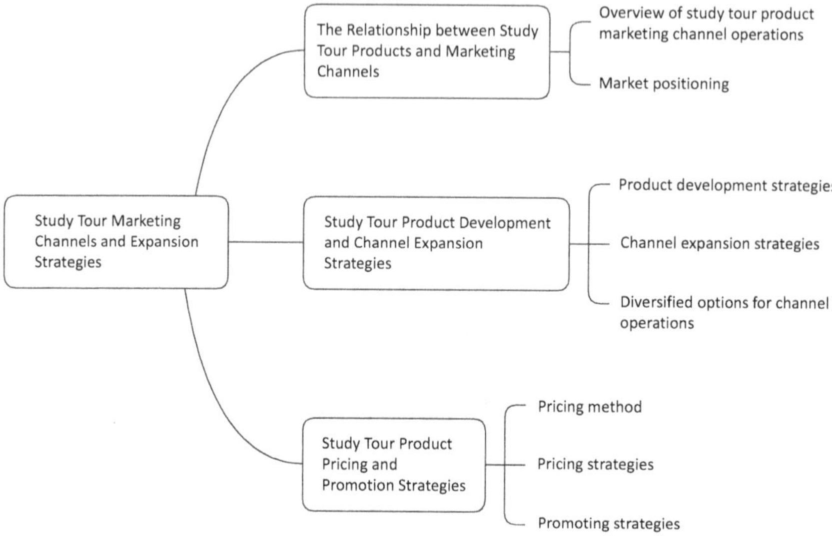
Study Tour Marketing Channels and Expansion Strategies
The Relationship between Study Tour Products and Marketing Channels
Overview of study tour product marketing channel operations
Market positioning
Study Tour Product Development and Channel Expansion Strategies
Product development strategies
Channel expansion strategies
Diversified options for channel operations
Study Tour Product Pricing and Promotion Strategies
Pricing method
Pricing strategies
Promoting strategies

Contents

1. Case Study

Diversified Content Drives Diversified Channels

As the study-tour industry moves towards greater diversification, a clear shift has occurred: the market is no longer centred on the traditional "spring-and-autumn outings" for schools. Content that enables students to learn while travelling and travel while learning has become the primary driving force. Both educational institutions and study-tour providers invest in high-immersion, experiential offerings to attract users and distribution partners.

Take Moruo Drama Town in Leshan, Sichuan, as an example. Designated as a model site under the city's "Le-You-Jia-Xue" brand, it leverages the literary stature of Guo Moruo to establish its IP. It has successively developed signature courses such as "Hundred Surnames· Family Precepts", "Little Picture Book · Big Drama", and "Heavenly Street Market". By weaving family traditions, drama culture, live performances, and local history into compelling experiential settings, these courses—each with a clear theme, storyline, tasks, real-world scenes, and hands-on components—have attracted numerous primary and secondary schools, parent-child tour operators, and OTAs. The site has been recognised as a Study-Tour Practice Base for schoolchildren in Leshan. On and off the stage, its rich, high-impact experiences have gone down well with parents and students alike, making it the top choice for distributors and family educational tourisms across Sichuan.

1.1. *Reflection*

1. Briefly discuss your understanding of the relationship between study-tour courses and distributors.
2. Share a case where a distributor marketed a study-tour product, and describe the detail that impressed you most.

Educational tourisms remain an emerging sector whose marketing channels are still developing. Because these tours are inherently experiential, it is difficult to find comprehensive channel information online; even when scattered details are available, they lack vivid, scenario-based appeal. Much of the information still spreads by word of mouth among parents. Surveys show that parents prefer to obtain study-tour information online to access itineraries and course details more fully and proactively. However, most providers' product pages are rudimentary and do not yet use video or other immersive media to convey the real experience.

2. The Relationship between Educational Tourism Products and Marketing Channels

2.1. *Overview of marketing channels for educational tourism products*

(1) The Concept of Educational tourism Product Distribution
Distribution of educational tourism products refers broadly to the pathways through which educational tourism service providers deliver their products to end consumers.

(2) The Concept of Marketing Channels for Educational tourism Products
Marketing channels for educational tourism products refer to the intermediary links through which educational tourism products are transferred from the developer to the end consumer. This also encompasses the number and geographic distribution of sales outlets within the channel network.

(3) The Necessity of Building Marketing Channels
To achieve sustainable growth, educational tourism service providers must increase product sales and improve service quality. Relying solely on a single sales channel is insufficient. Integrating and expanding multiple distribution channels is essential to build a healthy cycle between product supply and service delivery.

(4) The Functions of Marketing Channels

i. To promote educational tourism institutions' brand, products, and service quality, conduct market research through market linkage, and gather information about similar products or competing brands.
ii. To develop and iterate educational tourism products based on consumer preferences reflected through channel feedback.
iii. To help educational tourism service providers expand their customer base, maintain client relationships jointly, and serve consumers effectively.

2.2. *Market positioning*

(1) Intra-city Market Positioning

i. Target Consumers in the Educational tourism Market

According to the Ministry of Education's guidelines on experiential learning, the intra-city educational tourism market is segmented by school level:

- Primary schools emphasise local culture and community-based education.
- Lower secondary schools concentrate on themes at the county and city levels.
- Upper secondary schools concentrate on provincial and national-level themes. Educational tourism providers must organise courses to develop a coherent curriculum aligned with these requirements.

ii. Target Consumers in the Parent–Child Educational tourism Market

The parent–child educational tourism market differs from the general educational tourism market. Its primary audience comprises students in Years 1 to 3 of primary school, an age group with relatively limited independence that necessitates parental accompaniment. Typically born in the 1980s or 1990s, parents often possess higher educational backgrounds and significantly emphasise parent–child education. This has driven the demand for educationally oriented family tours, resulting in the emergence of the parent–child educational tourism segment.

iii. Courses and Curriculum Systems

Educational tourism courses should be systematised to cater to the needs of various age groups and destinations. They should be designed under the Ministry of Education's guidance and connected to age-specific learning outcomes and assessment standards, ensuring alignment with school-based educational tourism implementation.

iv. Core Competitiveness

In line with the K-12 framework, each educational tourism base should develop in-depth, age-specific course content (excluding

Year 9 and Year 12, which focus on examinations). Schools generally choose 10–20 educational tourism bases, each tailored to meet the needs of a specific age group or school year. The aim is to establish each base as the premier destination for delivering the curriculum for that age group, thereby enhancing core competitiveness in both the range of destinations and the depth of experiential content.

2.3. *Intra-provincial market positioning*

1. **Target Consumers**
   The intra-provincial educational tourism market mainly targets upper primary and lower secondary students (Years 5–8).
2. **Courses and Curriculum Systems**
   Although educational tourisms are ideally designed following the school year, in practice, they are often organised at the school level. Therefore, educational tourism destinations should be able to offer multi-grade course systems, making them more appealing to schools with diverse student populations.
3. **Core Competitiveness**
   Develop a unified K-12 curriculum system to establish core competitiveness. Different cities across the province provide varying educational tourism resources. With access to superior resources, national and provincial-level bases become key destinations. From a school management perspective, schools in smaller cities favour educational tourism bases that can accommodate all age groups in one location. Destinations must create comprehensive curriculum systems following the K-12 framework.

2.4. *Inter-provincial market positioning*

1. **Target Consumers**
   The inter-provincial educational tourism market primarily targets upper secondary school students, typically from regions with strong academic performance. However, many lower secondary schools also organise cross-provincial tours. Such activities can begin as early as

Year 5 of primary school in more developed regions. Nevertheless, the market is generally centred around junior and senior secondary students.

2. **Courses and Curriculum Systems**
   Cross-provincial educational tourism programmes must emphasise local characteristics. Given the wealth of domestic educational tourism destinations, only those with national recognition and distinctive experiential curricula will likely attract students from other provinces. Considering the higher costs and management complexity of cross-provincial travel, differentiation, compelling scenarios, and impactful content are key competitive advantages.
3. **Core Competitiveness**
   The priority is to offer distinctive educational resources to build a themed K-12 curriculum system. For instance, Yunnan Xinhua Educational tourism has developed ethnic culture classrooms for younger primary students. They organise local ethnic minority representatives to bring cultural lessons directly to educational tourism bases, where they provide hands-on ethnic education to upper primary students. For older students, such as those in lower and upper secondary school, the programme arranges in-depth "cultural roots" educational tourisms in remote ethnic villages in the mountainous regions of Yunnan.

   Secondly, the experiential content must be rich and varied. For example, Dayou Educational tourism Education Co., Ltd. in Anyuan District leverages the resources of the Central Party School to offer moral education that is both authentic and professionally structured. They have established a campus-based "Red Scarf National Flag Guard" and collaborate with cinemas, cultural venues, educational tourism bases, and children's centres to carry out "Little Red Lessons"—film-based moral education educational tourism programmes. They also organise the selection of outstanding Young Pioneers to attend the Red Scarf Cultural Roots Base in Anyuan, promoting patriotic education from an early age. These activities are deeply integrated into students' lives and classroom learning, contributing to a well-rounded moral, intellectual, physical, aesthetic, and labour development education.

Industry Case Study

Table 5.1. Intra-city One-day Educational tourism Products.

No	Recommended Activity Base	Recommended Course	Course Features	Suitable Grade Level
1.	Xingyuan Hill Shangxing Safety Education Base	Safety and Self-care	Enhance safety awareness, develop self-protection skills, and promote holistic growth.	Grade 1
2.	Douxianghua City Area	Little Photographer	Foster innovation and business awareness through photography-based experiential learning.	Grade 2
3.	Xingyuan Agricultural Experience Village	I Am a Little Farmer	Understand labour, divide tasks, love work, and strengthen life experience.	Grade 3
4.	Nickel Ore Museum	From Horses to High-Speed Rail	Understand transportation culture, appreciate its evolution and educational value.	Grade 4
5.	Dolphin Bookstore	Handmade Eternal Flowers	Connect with nature, identify flowers, and enhance hands-on and teamwork abilities.	Grade 5
6.	New Youth Workshop	DIY Handmade Soap	Students customise experiences and show creativity in handmade activities.	Grade 6
7.	Daxiahe Fossil Zone	Explore Fossil Crates	Discover 70-million-year-old fossil mysteries through field exploration.	Grade 7
8.	Anyuan District Red Education Base	Red Scarf Long March Experience	Walk the Red Route, feel a revolutionary spirit, and uphold ideals and beliefs.	Grade 8
9.	Dazhuang Maintenance Workshop	Little Repairman	Understand maintenance techniques, appreciate industrial beauty.	Grade 10
10.	Taihe Architectural Craft Base	Four Great Civilisations	Experience world civilisations, learn aesthetics, and understand architectural arts.	Grade 11

Provincial Multi-Day Thematic Educational Tourism Products

No.	Product Name	Course Highlights	Suitable Grade Level
1.	Liangzhu Ancient Civilisation (5-Day Tour)	Explore Liangzhu architecture and ancient technological development: "Cultural Symbols of Liangzhu" — learn wisdom of the ancient Chinese through elephant tusks, jade, sun worship, divine sacrifices, and ceremonial artefacts.	Grade 4–6
2.	Western Sichuan Agricultural Culture (6-Day Tour)	Explore Tibetan pastoral regions; intensely experience traditional Western Sichuan agricultural culture.	Grade 7–8
3.	Hongtong Clan Root Culture (7-Day Tour)	Trace ancestral roots and immerse in traditional clan root-seeking culture.	Grade 10–11

Educational Tourism Products for Other Provinces

No.	Product Name	Course Highlights	Suitable Grade Level
1.	Primary Student Winter Educational tourism: Western Sichuan Ethnic Culture (6-Day Tour)	Enter the Tibetan area and experience its unique ethnic customs: visit local ethnic villages, experience rural life, and immerse yourself in Tibetan traditions.	Lower primary & middle primary
2.	Junior High Winter Educational tourism: Enlightenment in Dunhuang & Hongtong (6-Day Tour)	Travel along the Silk Road and Dunhuang; appreciate cultural beauty through art education, cultural relics, and historical stories.	Junior high
3.	Senior High Winter Educational tourism: Enlightenment in Dunhuang & Hongtong (6-Day Tour)	Explore national cultural roots and moral heritage; learn about the Mogao Grottoes and Dunhuang mural stories; enjoy "FUN learning" and embark on a limitless cultural journey.	Senior high

3. Strategies for Developing Educational Tourism Products and Expanding Marketing Channels

3.1. *Educational tourism product development strategies*

Service providers must consider product differentiation and content depth when planning marketing channels for educational tourism products. Consequently, strategies for product development and expanding marketing channels will differ depending on the geographical scope.

(1) Product Development Strategies within the City
   1. Urban Products (Courses)
      Integrate the K12 curriculum with parent-child educational tourism courses. As an additional product purchased directly by parents, the parent-child educational tourism courses should emphasise engaging experiential content and high-quality experiential settings. This will gradually impact schools and parents' selection of educational tourism products.
   2. Spring and Autumn Educational tourism Products
      Traditional spring and autumn outings create a mature market. If these conventional outings' content and educational attributes are moderately upgraded, they can be transformed into educational tourism products. These enhanced offerings, priced slightly higher than traditional outings, possess significant market potential. Institutions newly entering the educational tourism sector should target this market by integrating traditional destination resources and developing products that fulfil market demands. These can subsequently be progressively enhanced into multi-destination K12 curriculum systems, with the core product aiding in the consolidation of market share.

(2) Provincial Product Development Strategies
   1. Provincial Products (Courses)
      Provincial products should emphasise diversity and breadth. A single provider often cannot fulfil all the demands of schools. Therefore, efforts should focus on mapping excellent existing courses available in the market and, through educational tourism associations or internal teams, developing a diverse curriculum system to meet the needs of schools. Note that the core of product diversification lies in curriculum integration.

2. Product Specialisation
   Due to the dual role of educational tourism providers—as both organisers and hosts—institutions with strong curriculum development and event management capabilities are more likely to gain school recognition. Product development teams should deeply research the characteristics of various provincial resources and design courses aligned with the curriculum, forming complementary and reinforcing relationships with school education, thus establishing a coherent curriculum system.

(3) Inter-Provincial Product Development Strategies

1. Out-of-Province Products (Courses)
   Inter-provincial educational tourism products typically feature higher unit prices and profit margins, making it the most competitive market. Schools assess destinations according to course uniqueness, hosting professionalism, and cost-effectiveness. Without a distinctive offering, it is challenging to prosper in this market.
2. Develop Products with Competitive Perspective
   The development of educational tourisms in China varies significantly by region. Some destinations have matured early, while others lack structured development. Therefore, course design must take into account nationwide competitiveness. For instance, Karst-related courses receive support from national-level resources in Chongqing, Guizhou, Hunan, Guangxi, and Yunnan. Providers must thoroughly prepare and identify core resource points to gain national recognition and distinguish themselves.
3. Parent-Child Educational tourism Products Outside the Province Represent a New Market
   Parent-child tours are priced higher and offer better margins, making them a competitive focus for educational institutions, parent-child service providers, educational tourism agencies, and travel agencies. This target group is mid- to high-end and demands much higher standards than schools. Therefore, specialisation is key; brand strength, renowned instructors, and rich experiential content are essential. Traditional high-quality service has become the baseline expectation. Consequently, without core resources, it is unwise to enter the inter-provincial parent-child educational tourism market lightly.

Model Case: Yunnan Xinhua Educational Tourism – Featured Curriculum Design

Yunnan Xinhua Educational tourism has chosen "ethnic culture" as its core theme in curriculum development. Yunnan is home to approximately 25 ethnic minorities, around 15 unique to the province. These rich cultural and human resources position Yunnan among the leading regions in China for ethnic diversity. As a major tourism destination, Yunnan offers stunning natural landscapes; however, its ethnic culture, an essential component of high-end educational tourism programmes, remains underdeveloped, presenting significant potential for growth. Furthermore, the Yunnan Xinhua Educational tourism team comprises seasoned professionals with over a decade of practical experience in ethnic cultural tourism.

Considering these factors, the Yunnan Xinhua Educational tourism conducted comprehensive research across eight key cultural dimensions: history, cuisine, architecture, clothing, festivals, marriage customs, language and script, and traditional arts. Based on this research, they developed a systematic and scientific curriculum framework divided into three levels. Firstly, they established a school-based curriculum by creating ethnic culture classrooms, designing a six-semester programme comprising 48 lessons, producing supporting teaching materials, and organising teacher training to integrate ethnic culture into the school. Secondly, they organised practical learning activities at educational tourism bases, where primary and secondary school students could engage in hands-on courses in various ethnic art forms. Lastly, they developed destination-based root culture educational tourisms focusing on immersive experiences in ethnic villages, such as those of the Dai, Yi, and Hani peoples. These programmes have facilitated cross-regional exchanges with other minority schools and have been widely embraced by students and teachers.

3.2. *Strategies for expanding educational tourism marketing channels*

(1) Development of Intra-city Marketing Channels

1. Competition in the Intra-city Market

   Brand recognition and reputation are vital assets. The school market is a typical "red ocean" market with intense competition. Its educational nature makes switching suppliers costly and complex. Therefore, establishing strong relationships directly with schools through high-quality courses (products) and dependable safety and service assurances is essential for maintaining long-term partnerships.

2. Building Educational and Sports Authorities into Marketing Channels

   Educational tourism enterprises can leverage their core competency—a multi-destination K12 curriculum system—to establish collaborations with local education and sports bureaus. These authorities can then promote the programmes to all primary and secondary schools under their jurisdiction. This is regarded as one of the most effective methods for expanding regional marketing channels in the educational tourism sector.

3. Developing Educational Tourism Bases into Marketing Channels

   Educational tourism companies can establish in-depth collaborations with educational tourism bases (camps) to develop single-destination K12 curriculum systems jointly. Together, they can operate and promote these offerings to local education authorities. The goal is to be included in the official list of recommended educational tourism bases, thereby gaining a first-choice advantage. Therefore, selecting distinctive bases and developing high-quality courses is a key direction for establishing marketing channels through educational tourism bases.

(2) Development of Provincial Marketing Channels

1. **Establishing In-house Channels**
   Many large educational tourism organisations have branches in various cities across their respective provinces. These branches operate under a unified curriculum and standardised implementation protocols to enhance brand credibility and gain market share.
2. **Provincial Alliances of Primary and Secondary Educational tourism Institutions**
   Educational tourism institutions typically serve both as organisers and hosts. However, expanding operations across an entire province is challenging and costly. As a result, alliances of educational tourism providers with core regional strengths have emerged—these are often formalised as provincial or municipal educational tourism associations. High-quality alliances establish

curriculum systems with province-wide applicability, then assign member institutions to deliver different course segments. Under centralised coordination and management by the association, members complement each other's strengths, share the market, and collectively build interconnected and advantageous marketing channels.

(3) Development of Inter-provincial Marketing Channels

1. **Exhibitions and Industry Familiarisation Tours as Catalysts for Marketing**
   Although offline exhibitions require investment in event promotion, they can yield substantial brand visibility. Advertising remains a vital promotional tool for educational tourism providers and may encompass brochures, print media, targeted content distribution, online ads, and self-media (social media) marketing.
2. **Educational Tourism Industry Seminars**
   Hosting industry seminars can enhance a company's influence in the sector and offer an opportunity to market to key target audiences directly. However, careful thought must be given to designing and delivering promotional content during such events.
3. **Joint Conference Platforms for Provincial/Municipal Educational Tourism Associations**
   Most members of these joint platforms are educational tourism associations at the provincial or municipal level, whose members are typically local educational tourism enterprises. These joint conferences provide participating institutions with robust support in policy advocacy and marketing channels, making them a high-quality and effective promotional avenue. Furthermore, associations fulfil a representative role in the industry and significantly influence the establishment of industry norms and standards.

3.3. *Diversified selection of marketing channels*

Integrating a range of diversified marketing channels to target different consumer groups with tailored promotions is essential. This approach helps to build a strong brand image for educational tourisms and creates immersive educational tourism experiences. Schools and industry stakeholders primarily represent the B2B (business-to-business) market, while parent-child educational tourisms target the C-end (consumer) market. Therefore, C-end marketing should be prioritised, focusing on digital promotion strategies. Online Travel Agency (OTA) platforms are critical for customer traffic, especially from the local market. Additionally, OTAs are increasingly embedded in local lifestyle service scenarios, offering convenient online booking options for local educational tourism products.

(1) C-End Marketing Channels

1. **Live Streaming Marketing**
   Live streaming marketing involves promoting educational tourism products through live broadcasts. These broadcasts act as platforms to enhance brand awareness and drive sales conversions for educational tourism providers. Although the conversion and monetisation of traffic via live streaming can be relatively slow, this method is particularly well suited to IP-driven parent-child educational tourism products that gain advantage from storytelling and emotional appeal.
2. **Short Video Marketing**
   Short video marketing is a form of content marketing that delivers valuable material to targeted users through brief videos. It aims to increase awareness of the provider's brand, products, and services and drive transactions. The key to success is identifying the right target audience and creating engaging content. With access to high-quality destinations and well-designed course offerings, short video marketing is a cost-effective and scalable tool for promoting local educational tourism offerings to a national audience.
3. **Community-Based Marketing**
   Community-based marketing ("social group marketing") has evolved from online community and social media marketing into a more intimate user connection and engagement model. It focuses on communication and building relationships to create user value. This humanised marketing strategy is popular among users and facilitates sustained word-of-mouth promotion. WeChat group marketing is a typical example and is particularly suitable for C-end parent-child educational tourism promotion within local markets.

4. **Building a New Media Matrix**
A new media matrix involves establishing a proprietary digital ecosystem for brand communication. This environment can showcase service standards, signature courses, experiential scenarios, teaching teams, and course reviews. Such a matrix enables consumers to access relevant information about educational tourisms and fosters trust and brand loyalty between the provider and its target audience.

(2) B-End Marketing Channels

1. **Institutional Visits and Tendering**
One of the key B-end marketing strategies is to conduct direct visits to schools and engage in tendering processes. Tender documents typically encompass company qualifications, corporate profiles, curriculum frameworks, teaching teams, proposed educational tourism products, contingency plans, standards and pricing, safety assurance systems, and evaluation frameworks.
2. **Cooperation with Local Education and Sports Authorities**
Educational tourism companies can quickly gain a competitive edge by partnering with local education and sports authorities to recruit student participants. This requires the company to have a well-developed curriculum system or maintain direct operational control over educational tourism bases (camps).
3. **Partnerships with Educational Tourism Service Providers**
Educational tourism institutions may collaborate with established service providers to create mutual value. The current market features numerous reputable providers such as Tianli Education and Zhongkai Travel Agency. Institutions should seek out like-minded partners whose products and markets complement their own. Partners can share course content, marketing channels, and co-branded initiatives through strong alliances, accelerating customer acquisition and enhancing brand exposure.

(3) Other Channels

1. **Educational tourism Industry Associations**
Active involvement in educational tourism industry associations is vital for provincial-level marketing. Participating in specialised educational exchanges and comprehensive industry conferences helps attract decision-makers from education bureaus, school

principals, teachers, and institutional stakeholders. Supported by quality content and the association's overarching marketing efforts, such engagement enables enterprises to showcase their competitive strengths, gain institutional trust, and achieve broader marketing objectives.

For instance, a travel agency in Yunnan partnered with the Yunnan Province Travel Agency Association's Educational tourism Committee, contributing to developing destination databases, curriculum repositories, and talent pools. It regularly participates in regional base certifications, curriculum promotions, and training programmes for educational tourism instructors. Through such initiatives, the agency has fostered strong, mutually beneficial relationships with local educational tourism institutions, facilitating student referrals and marketing expansion.

2. **Traditional Media and Public Relations**
Using traditional media and public relations encompasses branding efforts that broaden market influence. These efforts include appearances on television programmes, the production of documentary and promotional films, the publication of books, and the creation of advertising campaigns. Companies can enhance brand recognition across various social segments by animating static educational tourism resources and crafting engaging storytelling. Additionally, launching "Educational tourisms into Schools" promotional campaigns helps to gain the trust and endorsement of schools, parents, and students—key stakeholders in educational tourism.

4. Pricing and Promotion of Educational Tourism Products

4.1. *Pricing strategies*

1. **Competitive Market Pricing Strategy**
Educational tourism service providers incur relatively high operational costs in mature, large-scale, and highly competitive markets. Consequently, pricing strategies are generally competition-oriented, often leading to cost-based pricing. In such environments, small-scale providers may struggle to manage costs and exit the market due to

losses; mid-sized companies may dominate specific sub-markets, while large providers can leverage bulk purchasing advantages to secure a significant market share.

2. **"Blue Ocean" Market Pricing Strategy**
   Given the fierce competition and homogenisation of existing educational tourism products, there is a need to explore untapped "blue ocean" markets and develop new sub-market products. In such instances, a demand-based pricing strategy is employed. Its key advantage is that differentiated products can be priced higher, facilitating rapid cost recovery while maximising profits. However, popular products may swiftly attract competitors. Once the initial development costs are recovered, the pricing strategy should be adjusted to "cost + reasonable profit" to push competitors out and ensure long-term sub-market stability.

4.2. *Pricing methods*

1. **Cost-Based Pricing Method**
   This method involves setting the price as "cost + reasonable profit". Costs include base fees, course development fees, staff costs (including educational tourism instructors, safety personnel, and pastoral care staff), meals, transportation, insurance, materials, and accommodation for multi-day tours. Additional policy-related costs, such as expenses for two accompanying teachers per class or waived fees for a parent committee representative, may apply. As government support for educational tourisms varies by region, such variables must be carefully considered and appropriately adjusted in pricing plans.
2. **Product-Driven Pricing Method**
   A successful pricing strategy must consider consumer demand and competitor pricing. The price should cover production costs and target profit margins while persuading consumers that they receive exceptional value at the proposed price.
3. **Competition-Oriented Pricing Method**
   This approach prioritises gaining market share, even if it entails short-term losses. Given the lengthy and intricate value chain of educational tourisms and the high upfront investment, achieving economies of scale is essential. In numerous cities, it is common for a single

institution to dominate a district's educational tourism market. In such instances, referencing guidance prices from educational authorities and employing a cost-based model, typically with a 10% profit margin to uphold public service attributes, can ensure sustainable profits. Efficiency and cost control are crucial to enhancing profit margins.

4.3. *Promotion strategies*

1. **Local Promotions**
   1. **Organise Large-Scale Weekend Educational tourism Activities**
      Providers can showcase high-quality services and signature courses by regularly hosting weekend activities and inviting school leaders and extracurricular coordinators to visit the base. This exposure assists in attracting student groups through a cost-based pricing model.
   2. **Host Industry Conferences**
      Hosting education authorities, associations, and industry stakeholders at the educational tourism base helps to generate promotional impact. During the conference, providers can launch special offers and new product promotions, significantly boosting sales performance.
   3. **Parent-Child Educational Tourism Promotions**
      Arrange small contests or offer complimentary experience vouchers of symbolic value during activities. By providing engaging content, providers can attract parents and children, turning them into regular weekend parent-child educational tourism customers and achieving broader marketing goals.

2. **Intra-Provincial Promotions**
   1. **Promote at Professional Educational Tourism Conferences**
      Provincial and municipal associations often compile and categorise local educational tourism bases and courses. They organise industry expos at various locations, where providers can set up booths and offer promotional pre-sales, resulting in strong marketing outcomes.
   2. **Use Professional Promotion Platforms**
      Provincial associations' official course and resource databases can include high-quality courses and destinations. These offerings can

then be showcased on official platforms and marketed through pre-sales or "buy one, get one" campaigns, providing effective promotional leverage.

3. **Cross-Provincial Promotions**
    1. **Micro-Base Co-Development**
       Providers can invite marketing channels to co-develop micro-educational tourism bases, establishing shared ownership structures. This promotes diversified content creation and introduces stable traffic sources from partners, contributing to long-term, stable marketing goals.
    2. **Online Promotions**
       Educational tourism companies can create booking portals through OTAs like Ctrip and Tuniu, or lifestyle apps and platforms such as Douyin, aimed at businesses targeting specific provinces. Promotions may involve low-profit pre-sales and group deals to enhance product visibility in other provinces and convert national traffic into customers.

5. Chapter Summary

This chapter has deepened the understanding of the characteristics of intra-city, intra-provincial, and inter-provincial marketing channels by exploring the relationship between educational tourism products and marketing channels. It also examined corresponding product strategies, pricing strategies, and approaches for expanding marketing channels. This forms a solid theoretical foundation for the study of subsequent chapters.

6. Post-chapter Discussion Questions

1. Examine the similarities and differences in marketing channel expansion strategies for educational tourism programmes within intra-city, intra-provincial, and inter-provincial markets.
2. Which courses (products) listed in the main text best meet consumer demand? Please provide reasons to support your choice.
3. Which market is more effective in generating large-scale customer acquisition: B2B (institutional) or C2C (consumer)? Justify your answer.

4. Summarise the key strategies for designing educational tourism products. Provide a relevant case example, including the design approach, pricing strategy, and online promotional methods.
5. Design an on-site exhibition plan for a provincial educational tourism conference, concentrating on showcasing educational tourism bases and course offerings. Detail the key content and presentation format.

Chapter 6

Educational Tourism Market Branding: Development and Brand Architecture Strategies

Learning Objectives

1. Master the fundamental concepts of educational tourism branding.
2. Understand the characteristics of educational tourism brands and methods of brand promotion.
3. Grasp the pathways for building a educational tourism brand.
4. Learn about educational tourism brands' public communication and market entry timing.
5. Master product branding strategies for educational tourism offerings.

Chapter Framework

1. **Basic Concepts of Brand Management and Educational Tourism Brand Connotation**
   - Basic concepts of brand management.
   - Connotation of educational tourism product brands.

2. **Development of Educational Tourism Brand Products and Brand Assets**
   - Development of educational tourism products.
   - Assets of educational tourism brands.

3. **In-depth Understanding of Educational Tourism Brand Core Value Communication Strategies**
    - Interpretation and analysis of the difference between internal and external communication of brands.
    - Analysis of user perception, expression logic, and brand knowledge.
    - Analysis of the match between user expectations and product/ service performance.
    - Extraction of brand value keywords.
    - Visualisation and presentation of brand core values.

4. **Educational Tourism Brand Communication Planning and Publicity Timing**
    - Educational tourism on brand communication planning.
    - Construction and strategy of the media matrix.
    - Strategies for cross-platform communication, content, and formats.
    - Key stages and focus areas of brand communication.
    - Strategies for responding to public opinion crises.

5. **From Educational Tourism Product Branding to Educational Tourism Brand Persona**
    - Developing a brand persona.
    - Establishing a brand identity system.
    - Creating a brand communication image.
    - Managing brand reputation.

6. **Educational Tourism Product Brand Strategy**
    - Differentiated positioning strategies for educational tourism products:
        - Strategies for product brand communication.
        - Strategies for product brand development.

Learning Focus

1. Understand the methods for extracting educational tourism brand values and the steps involved in competitive product analysis.

2. Master constructing and integrating a educational tourism brand into the consumer lifecycle.
3. Learn the strategic approaches to building a educational tourism product brand.

Contents

1. Case Study: "Youth Naadam"

"Youth Naadam" is a traditional competitive sports event that reflects the grasslands' cultural essence and the Mongolian people's spirit. Based on the national spirit advocated by the conventional Naadam, "Youth Naadam" has established four core brand values:

1. Reverence and respect for nature;
2. Emphasis on family and taking responsibility;
3. Openness, inclusiveness, and valuing friendship;
4. Confidence, courage, and admiration for heroes.

Centred around the psychological growth of children, "Youth Naadam" designs educational tourism activities with themes such as "bravery and responsibility", "broadening perspectives,", "filial piety and affection", and "the pursuit of knowledge". These programs aim to preserve and promote grassland cultural values, support young people's healthy and joyful development, and strengthen harmonious family interactions. As a result, it has become a well-known cultural educational tourism brand in Inner Mongolia.

1.1. *Reflective questions*

1. Does "Youth Naadam" qualify as a educational tourism market brand?
2. What educational considerations should be addressed when building a educational tourism brand?

2. Basic Concepts of Brand Management and the Fundamental Connotation of Educational Tourism Branding

2.1. *Basic concepts of brand management*

Douglas Daft, former CEO of The Coca-Cola Company, once said: "Even if all of Coca-Cola's factories around the world were destroyed by fire, as long as the brand remains, the company could rebuild all those factories overnight from the ashes". This statement underscores the idea that a brand is the soul sustaining a business, an invaluable asset. Especially in the era of rapid technological revolution, where innovations evolve swiftly and product homogenisation intensifies, a brand becomes crucial for standing out in commercial competition.

In the educational tourism industry, customisation, safety, educational value, and the divide between decision-makers and end users create challenges in attracting customers on the C-end (consumers). A viable approach is to begin with similar products (such as parent-child science tours, study abroad programmes, and themed enrichment camps) to gradually raise parental awareness of "education + tourism" products, thereby increasing conversion potential. Guiding parents through a progressive journey of understanding and purchasing helps generate brand recognition for educational tourisms, which is critically important.

There are numerous definitions of branding. Marketing expert Philip Kotler defines a brand as "a name, term, sign, symbol, or design—or a combination of them—used to identify the goods or services of one seller or group of sellers and to differentiate them from those of competitors". From the perspective of Chinese culture, the meaning of "品牌" (brand) is also rich and insightful. Deconstructing the characters provides deeper associations: "品" implies quality, comprising functionality, emotional appeal (taste), and moral integrity (respectability), while "牌" suggests identity, reputation, or symbol. Thus, in Chinese, "brand" is even more nuanced.

The concept of "brand" has continuously evolved from academic and practical perspectives. The first stage of brand recognition shifted from a mark to a symbolic representation. The second stage evolved from a trademark to a promise. The third stage further developed into a promotional tool that became a strategic asset. As the digital and intelligent era of branding unfolds, it is now essential to build immersive virtual marketing environments and create exclusive virtual characters to engage more deeply with users.

2.2. *Core meaning of educational tourism branding*

From a practical application perspective, the essence of a educational tourism brand lies in high-quality products coupled with robust user engagement. A brand stems from the functional attributes of a product; thus, it is essential to focus on enhancing the intrinsic value of the involved resources. As user needs continue to evolve, it is also crucial to consider market demand orientation. Products must be consistently upgraded to meet these changing demands—otherwise, there is a risk that they will no longer align with user expectations.

Therefore, a brand should evolve alongside the development of its cultural value, gradually forming brand IP (Intellectual Property). When given a personalised identity, a branded IP creates a comprehensive image, facilitating diverse marketing and sustainable commercial transformation. In other words, IP-based branding may become a universal passport for educational tourism enterprises or destinations to thrive across various domains.

For the cities that host educational tourism destinations, the brand must interpret and emphasise their local cultural characteristics, showcasing the unique personality and charm of the city's educational tourism offerings. Anchoring the brand in local distinctiveness also promotes cross-regional cooperation and resource sharing, enabling the establishment of a well-coordinated, interconnected regional alliance of educational tourism brands—ultimately enhancing the overall brand development of the host city.

3. Development of Educational Tourism Brand Products and Brand Equity

A brand is not necessarily a concept apart from a product; it can be seen as a collection of product systems. At the same time, a product system embodies a process of thematic categorisation, enabling users to recognise a brand within a specific niche market swiftly and to extend its core value beyond that segment, thus granting the product a competitive edge. Ultimately, a product is essentially a reason to purchase. Therefore, when developing and designing a product, the primary emphasis should be on strategically defining the reasons for purchase, making the product relatively more competitive and easier to promote and sell.

3.1. *Development of educational tourism brand products*

Each systematic collection of educational tourism products helps reinforce the educational tourism's brand identity. This enables consumers to form strong brand recognition quickly and allows the brand to extend its core value beyond its original niche, amplifying its overall impact. For example, the Cultural Promotion Department of the Dunhuang Academy developed a series of educational tourism products based on the cultural heritage of the Mogao Caves and the academic resources of the Academy. These products were designed from various experiential perspectives, such as history, religion, archaeology, art, conservation, literature, music, and dance. They include themed programmes like: "On-site Mural Restoration—Meet the Mogao Caves' Mural Conservators", "Dunhuang Memories—Ancient Characters Brought to Life from the Murals", "Silken Buddhist Kingdom—Sharing Dunhuang Stories with Cultural Ambassadors", "The Dream of Mogao—A Fusion of Modern Technology and Ancient Art", and "Discovering Dunhuang Colours—Creating Pigments from Ancient Murals".

Once the product system was established, a group of experts and scholars from the Dunhuang Academy formed the Mogao Classroom Mentor Team, which led to the launch of the "Mogao Classroom" educational tourism brand. These interdisciplinary, knowledge-rich, and highly professional programmes were initially promoted through the Dunhuang Academy's official WeChat account, featuring a monthly recruitment post. At first, each session attracted only a dozen participants, but within a year, every session was fully booked due to excellent word-of-mouth. The brand value of "Mogao Classroom" became widely recognised.

Moreover, "Mogao Classroom" embraces the teaching philosophy of "Understanding, Experience, Creativity, and Inheritance". It offers programmes for both adults and youths. The youth classes have gradually developed their IP brand—"Little Mogao Ambassadors"—integrating brand strategy with product development. The audience is no longer restricted to domestic participants; the launch of international sessions has further promoted Dunhuang culture globally and fostered cross-cultural exchange.

3.2. *Brand equity in educational tourisms*

Brand equity originates in the consumer's mind; this is the ultimate goal of systematic brand building. It reflects brand performance in the product market and brand valuation in financial markets. Generally, the level of brand equity serves as a key indicator of measuring a brand's overall strength.

Brand equity manifests in the increased value of tangible and intangible corporate assets. Tangible assets directly contribute to the company's revenue, including the price premium consumers are willing to pay for the brand. Intangible assets—typically the more significant of the two—are driven by consumer perceptions such as brand loyalty, brand awareness, perceived quality, and brand associations.

Continuous marketing and brand development investment is essential to maintain and grow brand equity. This encompasses creating brand-related technologies, innovative service systems, and engaging product experiences to enhance consumer loyalty and reputation.

In today's digital age, the relationship between consumers and brands increasingly manifests in digital form. Digital brand equity is cultivated through online marketing strategies. A brand can generate significant digital equity by fostering a sense of community identity and encouraging interactive relationships between the brand and users. For instance, Xiaomi harnesses community-based marketing to enhance user engagement. Similarly, emerging internet-native brands such as HeyTea, Genki Forest, and Three Squirrels have gained distinct competitive advantages through digital strategies, swiftly building valuable digital brand assets.

The educational tourism industry is also progressing towards digitalisation through tools such as digital learning kits, educational tourism websites, and audio-visual course bundles. These innovations enhance brand loyalty and generate active user traffic, allowing brands to engage users through multiple online touchpoints. By leveraging demand-driven customisation and differentiated experiences, educational tourism brands can optimise each link in the operational chain and continually strengthen their competitive advantage.

4. Methods for Refining the Core Value of a Educational Tourism Brand

The issue of refining a brand's core value fundamentally revolves around brand positioning and shaping brand personality. A brand's core value signifies its most unique and valuable essence. It serves as the foundational basis for brand positioning, while brand personality is the

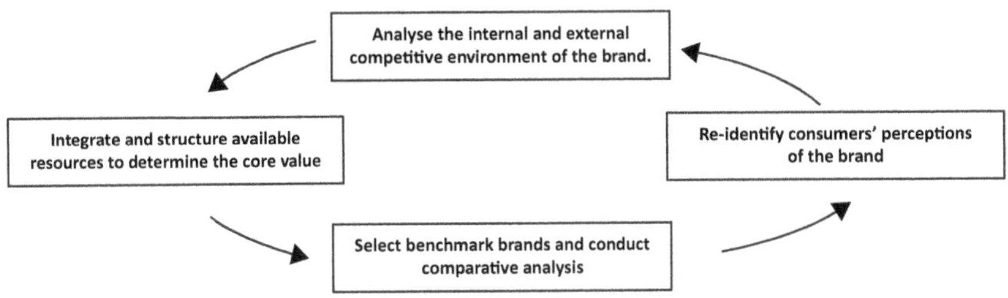

Figure 6.1. Methods for refining core brand value.

concentrated expression of this value. Positioning and personality must be determined based on the brand's core value, and these three elements must remain aligned and coherent. Therefore, refining a brand's core value becomes particularly important. Currently, methods for refining core brand value are evolving domestically and internationally. Drawing from practical experience, the author summarises four key approaches to refining core brand value, as shown in Figure 6.1.

4.1. *Investigate, analyse, and evaluate the internal and external competitive environment of the brand*

In identifying why the existing brand may require adjustment, a business should analyse the current market environment concerning market potential, competitors, consumer demands, and the company's strengths. It is essential to consider internal and external factors, particularly consumer perceptions and brand evaluations. SWOT analysis and Porter's Five Forces Model are widely used for analysing competitors.

4.2. *Analyse user needs and emphasise consumer perception of the brand*

Consumers' acceptance and understanding of brand culture is an interactive process. Through this interaction, consumers may gradually develop a positive attitude and emotional connection toward the brand or be influenced by negative perceptions and interpretations. In the era of big data, educational tourism enterprises can use insights into target consumers' need-based tags to gather numerous consumer opinions and creative

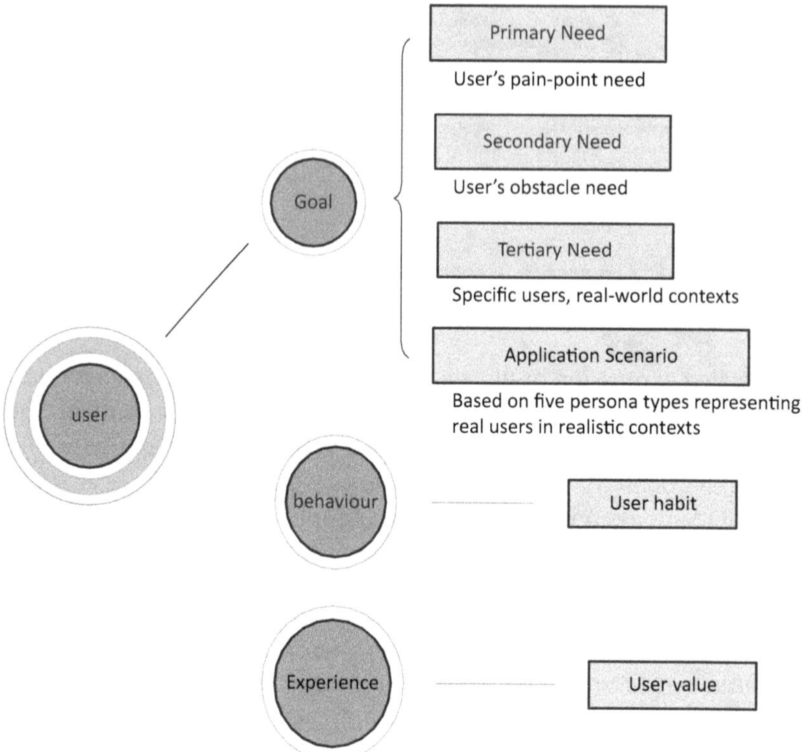

Figure 6.2. User demand tags from the perspective of the internet.

ideas. This aids in refining the brand's positioning by aligning it with consumer preferences, ultimately stimulating positive consumer responses and purchase intentions see (Figure 6.2).

4.3. *Methods and steps for analysing competitive brands in the educational tourism industry*

To comprehend a brand's positioning in the minds of learners and ascertain whether a educational tourism company can effectively showcase its advantages, or even leverage the recognition of competitors to refine its brand positioning, it is crucial to master the methods and steps of brand competitor analysis within the educational tourism sector.

(1) **Identifying Potential Brand Competitors**
Brand competitors refer to rival companies that offer similar products targeting the same customer group but operate under different brand names. The principles for identifying brand competitors include:

- **Market Share:** Identify the three leading brands based on market share.
- **User Feedback:** Gather insights into users' perceptions of each brand's strengths and weaknesses.
- **Company Background:** Larger enterprises often demonstrate greater resource integration capabilities and can expand swiftly. Following the implementation of the "Double Reduction" policy, numerous large educational training groups have quickly entered the educational tourism industry.

(2) **Determining Dimensions for Brand Competitor Analysis**

a. **Regional Competition Analysis**
Understand local brand competitors, compare their strengths and weaknesses, define your positioning, and develop corresponding marketing strategies.

b. **Product Perspective Analysis**
Concentrate on core educational tourism offerings, product ranges, and pricing structures.

c. **Consumer Analysis**
Includes:
   - Target customer analysis
   - Potential customer evaluation
   - Market size assessment
   - Analysis of the company's market position

d. **Sales Status Analysis**
Review essential financial metrics such as:
   - Operational costs
   - Marketing breakeven point
   - Product investment efficiency

4.4. *Summary report of brand competitor analysis*

A summary report should align with the original purpose of the competitor analysis; only then can it provide genuine value. The steps for composing the summary report are outlined in Figure 6.3.

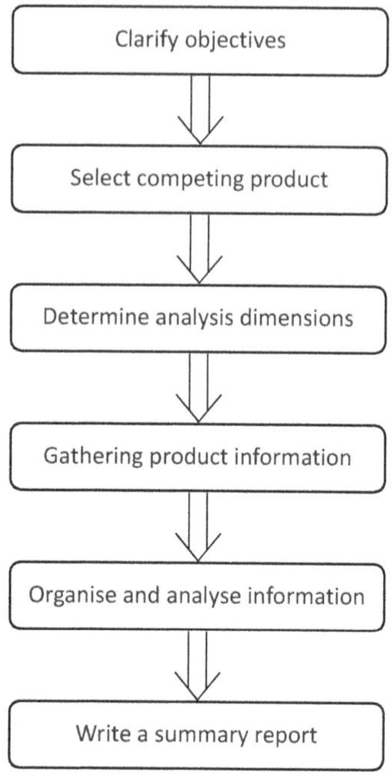

Figure 6.3. Steps to write a summary report.

1. Structuring Resources to Define Core Brand Value
   Educational tourisms represent the integration of tourism and education, aiming to achieve a synchronised resonance between both sectors. Unlike traditional classroom-based education, educational tourism products transcend textbooks, classrooms, and school settings, offering a wealth of content through immersive, real-world experiences. A brand is a complex system, and distilling the essence of a educational tourism brand requires close attention to the uniqueness of the available resources. Priority must be given to the educational attributes embedded within those resources and to identifying how well they align with experiential learning scenarios.

   This process involves structuring resources—extracting and refining key elements to ensure alignment with educational objectives—and establishing a differentiated brand value. The structured, distinctive resources must then be effectively paired with experiential goals. Moreover, when organising resources, it is crucial to consider how well they align with consumer needs, particularly those of primary and secondary school students. Resources should be adapted to correspond with students' psychological and developmental stages across different age groups, integrating a clear sense of progression.

When integrating family travel with educational tourisms, carefully managing the breadth and depth of brand refinement is crucial. Therefore, maximising the development of experiential settings is essential, enabling consumers to enjoy high-quality services. Exceptional resources and experiential environments serve as the "moat" that protects brand value and embody the core of the brand's unique identity.

Case Analysis

Youth Cultural and Museum Educational Tourism Brand—"Be an Archaeologist for a Day"

The Ancient Shu Civilisation Exploration Base, located 6 kilometres from the Sanxingdui Museum, is a tangible, immersive extension of the "Sanxingdui experience". The founding team of the base recognised that contemporary cultural and museum-based educational tourisms often lack variety and interactivity. Most activities depend heavily on teacher-led explanations and brochure reading, leading to low student engagement and limited attention retention. They realised that by transforming museum exhibits into teaching aids and incorporating games or competitive elements, they could considerably enhance student interest and motivation.

To this end, the base replicated Sanxingdui artefacts on a 1:1 scale and reconstructed the No. 2 sacrificial pit based on archaeological excavation reports. With this foundation, they designed a series of archaeology-themed courses, including:

- Simulated excavation of the Sanxingdui site.
- Ancient Shu craftsmanship: bronze head casting using the lost-wax method.
- Ancient Shu craftsmanship: jade carving employing traditional techniques.
- Sounds of Sanxingdui: bronze bell ensemble performance.

These modules are all centred around the immersive project scenario titled "Be an Archaeologist for a Day", which authentically recreates the archaeological field environment. Students gain hands-on experience with excavation techniques, acquire specialised knowledge, and actively and meaningfully engage with cultural relics. Through this innovative approach, the base has successfully developed the "Be an Archaeologist for a Day" youth cultural and museum educational tourism brand.

Reflection and Discussion Prompts:

1. What experiential scenarios has the "Be an Archaeologist for a Day" project designed? Can you think of any further suggestions or enhancements? Please read the case thoroughly and share your reflections.
2. The "Be an Archaeologist for a Day" project benefits from strongly endorsing the Sanxingdui brand. Can you think of other educational tourism brands that successfully leverage core resources for brand credibility? What strategies can be employed to build brand value and trust if no such core resources are available?

5. Brand Building Pathways and Brand Introduction for Educational Tourisms

5.1. *Top-level planning of educational tourism brands*

The systematic construction of a brand follows a path guided by the overall development strategy of a destination or enterprise, creating a market-oriented branding framework based on the dual evaluation standard of "core resources + market demand". Its essence lies in establishing a brand marketing cycle centred around optimising user experience and fostering consumer brand loyalty.

"How to effectively carry out top-level brand planning" has become a key developmental bottleneck for destinations and enterprises. Educational tourisms represent the integration of "tourism + education + culture"—a harmonious resonance of education, tourism, and cultural elements. These products transcend the confines of textbooks, classrooms, and schools, offering rich learning experiences within real-world environments.

Refining a educational tourism brand requires emphasising the uniqueness of course resources from a study-focused perspective, identifying the brand, and establishing clear goals, pathways, steps, and implementation measures to build a competitive and sustainable brand. From a practical market perspective, the following six principles should be adhered to:

(1) **Goal-Oriented Principle**
Top-level brand planning must be target-driven. Enterprises or destinations should adopt a results-oriented approach: set clear, specific goals first, then work backwards to design implementation plans, considering both execution and strategy while taking current development status into account.

(2) **Long-Term Principle**
The Long-Term Principle emphasises that brand planning should align with the brand's lifecycle, necessitating different strategies from both the consumer and brand perspectives at various stages. Many educational tourism products encounter challenges, such as limited variety, weak targeting, and a failure to meet the needs of segmented markets. Some also struggle with poor standardisation and inconsistent quality. In the short term, clarifying the brand's development goals is essential. In contrast, in the long term, efforts should focus on diversifying communication channels to enhance brand visibility and ensure the timely iteration and optimisation of new products. Brand planning can maintain value and strategic relevance by addressing immediate and future needs.

(3) **Consistent Positioning Principle**
Brand value and market positioning should remain stable and not change arbitrarily. Brand positioning is built upon product and market positioning. Once the brand positioning is established, implementing market positioning becomes more straightforward. Positioning based on market demand helps the target audience form a clear mental association with the brand.

(4) **Normative Execution Principle**
Top-level brand planning must comply with brand standards to ensure consistent implementation. This relates to brand image, communication, and product design—each must adhere to standardised brand system rules.

(5) **Systematic Principle**
Brand building is a comprehensive task. Brand content development should have clearly defined audiences and a consistent tone, rather

than being managed through ad hoc arrangements. A systematic framework, transparent processes, evaluation goals, and regular assessments must be adopted to ensure that the output aligns with strategic brand positioning.

(6) **Broad Participation Principle**
Effective brand planning involves widespread participation. In addition to a structured development pathway, it should invite opinions from all stakeholders through incentive mechanisms and brainstorming sessions, both online and offline. Engaging the entire team increases their sense of ownership and encourages deeper involvement in constructing a coherent, strategic brand plan.

Case Analysis

"Leyo Jiaxue" Educational Tourism Brand Standard Development and Implementation

Leshan City has successfully developed its educational tourism brand "Leyo Jiaxue" with a thoughtfully designed brand name, logo, IP character, and cultural-creative products. However, the brand lacked a comprehensive system of standards and a structured implementation framework. To address this, the Leshan Municipal Bureau of Culture, Radio, Television and Tourism established a set of brand standards for "Leyo Jiaxue" following existing guidelines for educational tourism practice base facilities and service norms, product design protocols, as well as local resource characteristics, future development needs, and international benchmarks. These standards aim to guide the creation and promotion of the "Leyo Jiaxue" brand.

According to the *"Leshan City Primary and Secondary School 'Leyo Jiaxue' Educational tourism Implementation Plan"*, the city rolled out a brand rating system that standardises the evaluation procedures for bases, courses, instructors, and safety management, promoting high-quality development of educational tourisms across Leshan. Once the standard system was established, the bureau further refined strategies for brand promotion, storytelling, and communication, expanding the brand's reach and aiming to present Leshan's finest, most unique, and most joyful educational tourism offerings across Sichuan and globally.

Guiding Questions

1. What has the Leshan Municipal Bureau of Culture, Radio, Television and Tourism done in the top-level planning of the "Leyo Jiaxue" educational tourism brand?
2. Which principles did the planning and construction of the "Leyo Jiaxue" educational tourism brand follow?

5.2. *Establishing a educational tourism brand identity system*

Brand identity is often perceived as a distinctive visual symbol of a brand, such as a logo, a colour palette, or an emotional tone. Consumers frequently interpret brands through perceptions like "trendy", "comforting", or even mystical attributes. However, brand identity design should be holistic, systematic, and grounded in scientific logic within the actual business environment.

As experts and scholars from the University of the Arts London state, "Create a distinctive point, give the brand meaning, and convey a sense of trust". This definition encapsulates the essence of logical and systematic brand identity design. A widely adopted process for designing a brand identity system includes the following stages: clarifying the brand's value and concept, naming the brand, and developing the brand's logo and visual identity system.

(1) **Clarifying Brand Value and Concept**

A well-designed brand identity must address several stages: the concept stage, text stage, symbol stage, and system stage—corresponding to the brand's philosophy, brand name, brand logo, and overall brand system application. A comprehensive system is essential for expressing brand value; only with systematic design can a brand's added value become a compelling reason for consumers to purchase.

A product's original value stems from the producer's intention and the context of its production. Once it enters the market, its fundamental meaning and value must be organised and presented clearly. The essence of branding is to clarify and make this value accessible, aiding consumers in quickly understanding and emotionally connecting with it. This represents the primary role of brand identity design:

to minimise the communication cost between producers and consumers while satisfying the need for intuitive and immediate product engagement recognition.

Thus, defining a brand's value and concept establishes the foundation of any identity design. The emphasis is on uncovering the brand's uniqueness and differentiation. For instance, the designers undertook extensive field research for a previous branding project centred on a local speciality product from northern Shaanxi. They discovered that many areas in this region continue to uphold traditional farming practices, untouched by modern machinery or industrial pollution due to the challenging terrain. These pristine conditions resulted in high-quality agricultural products. In light of this, scholars introduced the concept of "natural farming" and developed the identity theme "Yellow Earth Style, Naturally Red", crafting a unique and culturally rooted brand identity system.

(2) **Brand Naming**

Once a brand's core concept and value system are established, its cultural essence takes shape. Brand culture represents an elevation of brand values—it adds a spiritual and emotional dimension to a brand, transforming a consumer's basic material need into a deeper emotional connection. This infusion of vitality makes the brand more relatable and easier for consumers to accept. A compelling brand story helps convey this cultural essence by identifying a representative event, object, figure, or visual that encapsulates the brand's core value and spirit. This narrative enhances cultural richness and allows consumers to recall and associate with the brand more easily when triggered by promotional messages, forming lasting impressions.

Therefore, brand naming is essentially about telling a good brand story. Names like "Three Squirrels" and "Bestore" (Liangpin Shop) easily spark the imagination and evoke potential stories behind the brand.

(3) **Creating the Brand Logo and Visual Identity System**

In earlier brand design, logos were primarily explained through symbolism and deeper meaning. The visual identity system (often encapsulated in a VI manual) merely indicated that a product or service belonged to a particular brand. However, with the strategic foundation discussed previously, the logo and visual design now emphasise social aesthetics and market trends. A strong brand visual identity must be distinct and easily distinguishable from competitors, allowing the target audience to recognise and remember it quickly. Maintaining visual consistency over time is crucial—frequent changes in branding diminish recognisability and weaken the consumer's ability to form a clear brand image.

The brand identity design process outlined above can assist most businesses in building an effective brand system. However, in today's fast-paced, information-driven era, branding encounters a new challenge: personalisation. At its core, developing a brand identity system addresses the issue of standardisation—it transforms abstract concepts into tangible, repeatable standards, much like the rise of chain stores. Yet, standardisation can sometimes feel "official" and distant, reducing flexibility in adapting to the diverse demands of a rapidly changing world. That's why the concept of **Brand IP** (intellectual property) has emerged.

A Brand IP embodies a character, a set of values, and specific paths for communication. As illustrated in Figures 6.4 and 6.5, this personalised expression allows the brand to resonate more deeply. In this context, "Brand IP Design" refers narrowly to a brand's effort to craft a personalised, character-driven identity. In recent years, many brands have embraced this approach. While the traditional visual identity belongs to the realm of standardisation, Brand IP offers a powerful vehicle for expressing individuality in today's branding landscape.

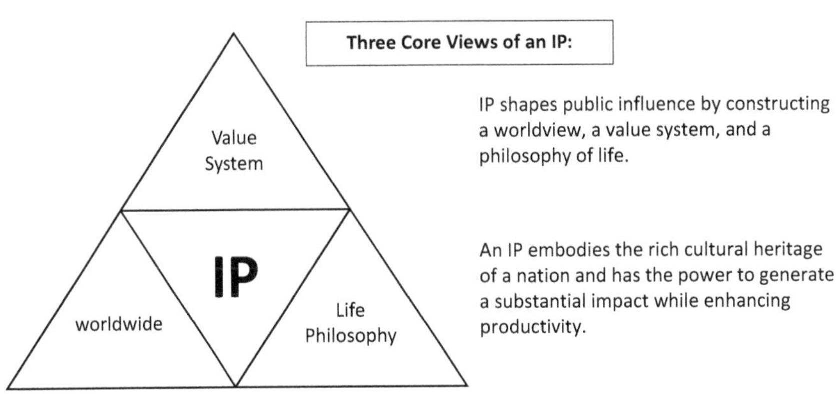

Figure 6.4. The three dimensions of IP.

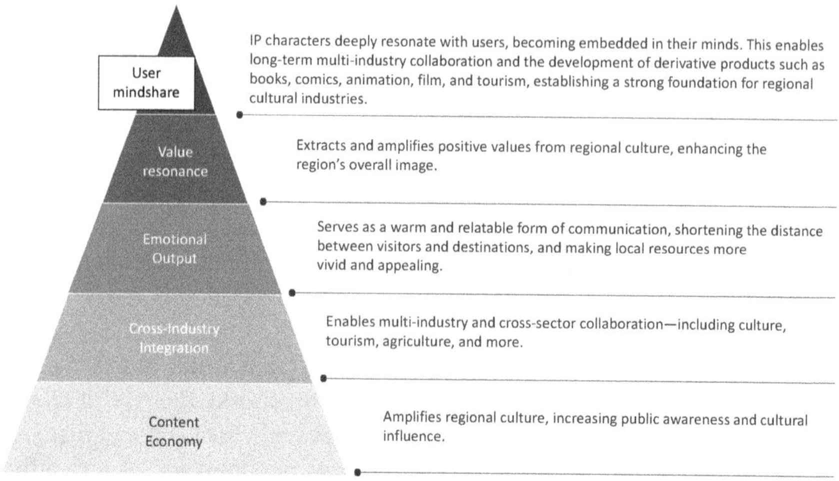

Figure 6.5. Path attributes of IP communication.

IP Attributes

Much like the notion of brand image design, whether it pertains to intellectual property associated with cultural content or a character developed through the IP-ification of a non-cultural brand, each emotionally resonant IP design is founded upon a clear cultural core.

So, how do we accurately identify the cultural essence behind an IP? This requires following a basic reasoning process: First, clearly list the brand's intrinsic values. Then, define a unique core entry point around which differentiation can be built—this step remains relatively abstract. To transform this abstract concept into something tangible and understandable for consumers, finding a real-world cultural or social phenomenon that can serve as a carrier is essential. This carrier should conceptually align with the brand's values.

For example, in the creation of the "Hip-Hop Warriors" IP, the designer drew inspiration from a traditional folk event popular in Suzhou called Nao Mengjiang (literally "Raising a Ruckus with the Warriors"). Initially, few people knew what Nao Mengjiang was, but upon further exploration, two key elements emerged: "Nao" (to cause a lively commotion) and "Mengjiang" (fierce warriors). The event itself resembles a large party—a vibrant, collective celebration. This concept closely aligns with hip-hop culture, which has gained massive popularity among young audiences in recent years due to its high visibility and viral appeal. Moreover, the energetic and rebellious atmosphere of hip-hop mirrors the spirited essence of Nao. The idea that "every master is a warrior" perfectly bridges the two, giving birth to the IP concept of "Hip-Hop Warriors". When warriors come together, the ruckus begins.

Brand image building is an ongoing process with no endpoint. As long as a brand exists, its image requires continual refinement. Even a well-established brand image may fade as consumer preferences and needs evolve. When a brand image becomes diluted or forgotten, it signifies a loss of brand vitality. Therefore, to maintain a consistent, clear, and enduring image in consumers' minds, brands must perpetually update their image and create new value.

Schools are encouraged to take the lead in establishing a comprehensive service framework for educational tourism programmes. This ensures that educational tourism offerings closely align with the curriculum and practical educational philosophies. With ongoing institutional support, standardised management, and cross-sector collaboration, educational tourisms can evolve into a regular, structured extension of classroom-based education.

5.3. *Educational tourism brand product system support*

The brand product system for educational tourisms consists of a collection of various product forms under the educational tourism umbrella. These different expressions are interconnected and represent the core value delivered to consumers. Systematising products involves classifying them by theme or category, enabling consumers to achieve quick brand recognition within specific market segments and gradually expand the brand's perceived value beyond those segments, ultimately enhancing the product's competitive advantage.

For instance, the product system of "Baby Walks the World", an outdoor education institution, positions itself as a leading provider of children's outdoor activities in China. Its system is supported by six thematic product lines: urban survival education, outdoor challenge education, natural history museum education, military character education, camp-based outdoor education, and international educational tourisms. These structured offerings assist parents in understanding the benefits of outdoor education for children, building survival skills and fostering scientific thinking through hands-on experiences. Moreover, this institution is the first in

China to obtain ISO9001 certification for "Domestic Children's Educational tourisms", which strengthens the safety framework of its experiential products and has earned the trust and support of many parents.

Furthermore, timely product development strategies must be taken into account, as they directly affect the success or decline of a product line. Adopting a "hit product" mindset is vital, particularly during the initial stages of product development. By concentrating on a flagship offering, a brand can quickly build awareness and reputation, thereby facilitating the promotion of its wider product range.

Take the educational tourism product from Chishui, Guizhou, as an example. It centres around the theme "Four Crossings of Chishui, Red Awakening of China", which successfully established the "Four Crossings of Chishui" as a signature red tourism educational tourism brand. For primary and secondary students, the core theme is "Tracing the Long March, Revisiting the Soul of Chishui". The thematic camp is aptly named "Red Growth Chronicles: Chishui Exploration Camp", strategically focusing on this historical motif to enhance recognition and impact.

5.4. *Support from the educational tourism brand service system*

Educational tourisms integrate "learning" and "travel", and the successful implementation of educational tourism brand products depends on a systematic service framework. Since educational tourism implementation involves multiple stakeholders—including government bodies, schools, educational tourism bases, agencies, and parents—establishing a mutually supportive service system is essential. Creating a school-led service system for academic educational tourisms and a base-led system for parent-child educational tourisms helps ensure that the courses are effectively implemented. Three key mechanisms underpin this framework:

(1) **Multi-Stakeholder Collaboration: "Government + Schools + Society"**
Government policies encourage schools to incorporate educational tourisms into their formal programmes and actively promote participation. However, schools often face constraints such as safety concerns, limited funding, and a lack of organisational experience

among regular teachers. At the same time, many external institutions struggle to combine "learning" with "travel effectively".

Therefore, schools need to lead the construction of a service system that ensures educational tourism products align with curricular and experiential education philosophies. Through continuous institutionalisation, standardisation, and multi-party collaboration, educational tourisms can be fully integrated into school-led, routine, extracurricular educational activities.

(2) **Rational and Effective Matching of Teaching Staff**

In educational tourisms, the base is the carrier, the curriculum is the core, and the mentor is the key. The matching and quality of educational tourism mentors have become critical issues. A educational tourism mentor is an emerging professional with tour guiding capabilities and educational qualifications, making them central to the delivery of educational tourism services.

Within a school-led service system, schools must establish educational tourism curriculum service teams to coordinate the implementation of experiential courses. These teams can collaborate with third-party institutions to conduct pre-departure safety training, plan emergency response strategies, and design safe outdoor learning activities. Simultaneously, balancing the school's academic curriculum with the educational tourism curriculum is vital—poor scheduling or overburdened teachers can adversely affect the programme.

(3) **Educational Tourism Base Support Focused on "Safety + Experience + Education"**

The base must integrate engaging experiential scenes with the course content as the central focus of educational tourisms. By creatively mining and utilising local resources, bases can offer immersive and interactive experiences. These activity-based learning methods help transform educational tourisms from passive, space-shifting lectures into meaningful educational engagements that enhance learning outcomes. From the perspective of infrastructure safety, factors such as on-site transportation safety, facility safety, fire safety, and food hygiene must be managed stringently. Providing high-quality services, including systematic reception procedures and professionally trained mentors, is a fundamental pillar of a comprehensive and reliable educational tourism brand system.

For destinations aiming to establish themselves as reputable educational tourism locations, leveraging leading enterprises as flagships is essential. These benchmark players can drive industry standards and consistently support the brand's successful implementation and long-term development.

5.5. *Support system for educational tourism brand communication*

With the rapid development of the internet, many educational tourism enterprises have begun to build their brand image through online promotion, primarily utilising new media platforms such as WeChat official accounts and short video platforms like Douyin (TikTok). Much of this content concentrates on current educational tourism policies and products, yet it often lacks professional management and strategic oversight.

A critical decision for these enterprises is establishing an in-house media operation or collaborating with third-party multimedia platforms. To make this decision effectively, it is essential to understand the framework and functioning of a brand communication support system and to evaluate the advantages and disadvantages of online services.

(1) **Brand Communication Must Be Purpose-Driven with a Strategic Plan**
Communication should be goal-oriented. Although the Return on Investment (ROI) for content marketing is challenging to quantify, it remains essential for management to define the purpose and position of brand communication clearly. From a branding perspective, sustained brand positioning and value promotion are crucial for fostering strong consumer recognition. Organisations often strive for their content efforts to align with business goals closely. Therefore, communication initiatives must be supported by a systematic strategy, clearly defined long-term and short-term objectives, explicit purposes behind the communication, and well-identified target audiences.

(2) **Produce Branded Content and Prioritise Media Credibility**
When producing branded content, it is essential to target the right audience by understanding their needs, challenges, and whether the content aligns with their expectations. However, many educational tourism enterprises often speak solely from their perspective,

neglecting audience preferences. To be effective, branded content must resonate with the audience's interests; otherwise, relying solely on fan counts or broad outreach is futile. Since educational tourisms involve schools, parents, and students, enterprises must consider the credibility of the media channel. Traditional media, such as television and newspapers, still retain significant authority and influence. Therefore, collaborating with reputable outlets, particularly those focused on education, can help promote the educational tourism brand across various educational levels.

(3) **Proactive Content Promotion Can Create Unexpected Opportunities**

The key challenge in content promotion lies in how to deliver it effectively. With limited budgets, many enterprises are better off focusing on social media platforms, particularly short video formats, and emphasising creativity in both content and presentation. For instance, in August 2021, locally produced MVs such as "Walking Tianfu Second Street Nine Times" and "Thank You for Your Dedication" captured the festive spirit in Chengdu and went viral through organic sharing. Hitting emotional and cultural touchpoints is an extremely effective way to achieve virality. For educational tourism brands, endorsements from niche experts can also be influential, whether through expert-curated posts or on-site experiences. Collaborating with influential content creators or WeChat public accounts for mutual promotion can enhance visibility. If opting for paid advertising, it is crucial to understand the latest platform algorithms and influencing factors and plan a promotion calendar that captures time-sensitive opportunities. Intense, timely promotion helps build consumer recognition and brand presence.

(4) **Evaluate the Effectiveness of Brand Content Distribution and Optimise the Communication System**

Optimising the brand content communication system involves refining the previous three stages—communication planning, content creation, and dissemination—based on performance evaluation.

Firstly, the effectiveness of brand content communication should align with the overall branding goals of the enterprise. From a value-creation perspective, basic metrics such as views and shares must be assessed, particularly by identifying which channels drive the highest traffic. It is also crucial to monitor the performance of brand-related keywords. Given the significance of brand reputation in educational tourism enterprises, tracking negative feedback and establishing a long-term mechanism for overseeing brand content distribution is essential.

Secondly, analysis should include evaluating the impact on sales conversions, lead generation, and customer retention. Data-driven insights can assist in refining the content strategy, focusing on metrics relevant to the target audience and specific market segments. A system for internal content assessment and external audience feedback should be established. Collecting and evaluating consumer feedback enhances the feasibility and relevance of the content, facilitating continuous improvement and more effective communication outcomes.

6. From Building a Educational Tourism Brand to Integrating the Consumer Life-Cycle

In his 1967 book Marketing Management, Philip Kotler stated that marketing is about managing consumer needs. The process of transitioning from brand construction to consumer life-cycle integration fundamentally involves dynamically managing consumers throughout. At various stages of brand development, analysing audience psychology and experiential feedback is essential to establish a comprehensive brand system progressively. This process is generally divided into three stages.

6.1. *Pre-market introduction phase*

During this phase, the primary focus is on making the brand image that embodies the core brand values visible and engaging to consumers, thereby initiating preliminary communication. The primary entry point is through product experience, supported by creating a professional mentor team and providing a high-quality service system. Initial launch events or experience tours help generate word of mouth within consumer circles, prompting users to pay attention to the brand. In this phase, integrating the product, visual identity, and communication systems is particularly crucial, as it establishes differentiated brand positioning and helps consumers understand “Who we are” and “What we do”.

6.2. *Brand growth phase*

At this stage, the focus shifts to product experience. The priority is to effectively implement resource allocation for service delivery, ensuring that the brand maintains market appeal through standardised and procedural service operations, robust safety management, and thorough product safety evaluations. Collecting and consolidating teaching and evaluation feedback fosters positive consumer recognition of the brand's value proposition. Through event-driven, periodic activities, continued consumer interaction sustains attention and builds purchasing confidence. The successful integration of the product and service systems becomes the core factor, helping consumers discern "What we offer" and "How we operate", thereby solidifying the brand's competitive advantage.

6.3. *Brand maturity phase*

In the maturity stage, the brand thrives by consistently delivering sustained, high-quality consumer experiences. However, the brand's vitality may decline under the constant pressure of increasingly homogeneous products. At this stage, it is crucial to strengthen consumer loyalty and enhance the brand's ability to command a market premium. Consistently fulfilling brand promises and avoiding negative publicity helps ensure that consumers maintain long-lasting, firm trust in the brand. It is essential to reward existing customers and encourage them to introduce new ones. For example, during Xiaomi's tenth anniversary, Lei Jun announced that the company would gift 1,999 yuan to each of Xiaomi's first 184,600 users as a token of gratitude—this prompted enthusiastic sharing among "Mi Fans" on social media and a strong sense of pride in being part of the brand.

7. Branding Strategies for Educational Tourism Products

7.1. *Educational tourism product line expansion strategies*

With the Chinese government continuously supporting the educational tourism sector, particularly following the Ministry of Education's "Double Reduction" policy in 2021, fresh momentum has been injected into the industry. The segmentation of "Educational tourism +" content has emerged as an increasingly important avenue for product development. According to Ctrip's 2021 Summer Tourism Big Data Report, leisure parent-child trips and educational tourism experiences dominated the customised travel market. Educational tourism product bookings increased by over 650% compared to summer 2020, and related search volume more than doubled. Over 80% of families with children searched for educational tourism-related travel products.

Despite this growth, many parents still lament that **"good products are hard to find".** The current difficulties in educational tourism product development can be summarised in four key issues:

1. **Too Many Counterfeits—Is It a Study or Just a Tour?**
   Most so-called educational tourisms remain fundamentally focused on tourism, offering little or no educational content— "more tour, less study" or even "just tour, no study".
2. **Lack of Substance—Straying from the Core Purpose**
   Many products lack contextual understanding. The essence of educational tourisms is learning, ideally through immersive, scenario-based teaching. The emphasis should be on curriculum experience and cognitive engagement—yet these intangible but essential aspects are often diluted or completely neglected absent.
3. **Homogeneity—Limited Variety, One-Size-Fits-All**
   The current offerings are generally limited in number and monotonous in style, characterised by rigid, lecture-based models with minimal interaction. Most rely on standardised delivery methods and similar equipment, simply altering the learning environment—this leads to considerable product homogenisation.
4. **Lack of Distinctiveness—Poor User Experience**
   Current educational tourism courses are fragmented and inadequately structured. They lack a tailored design for different audience segments and do not reflect deeper cultural dimensions (e.g. local, national, ethnic "root culture"). This results in derivative, copycat products, especially noticeable among educational tourism camp offerings.

Many educational tourism providers have begun diversifying into various product categories to tackle these developmental bottlenecks, aiming to capture markets within each segment. For example, one organisation has launched labour education tours, scenic educational tourisms, science exploration tours, and patriotic education tours.

However, most enterprises possess limited resources and cannot effectively cover all fronts. With widespread product overlap and increasing competition, market-oriented strategies focusing on well-defined niches—"small but refined"—typically outperform broad, all-encompassing product lines. This is especially critical in the post-pandemic era, where uncertainty and risk remain high. For small to mid-sized educational tourism businesses, concentrating limited resources on a few robust brand offerings is essential. A prime example is the Hong Kong-Zhuhai-Macao Bridge (HZMB), a landmark symbolising China's national strength. Integrating history and social studies, marine ecology, economics, and bridge engineering, the HZMB has become a valuable foundation for curriculum-aligned educational tourisms. In 2021, providers collaborated around the HZMB to launch a patriotic (red) series: the core product, Marine Study on the Hong Kong-Zhuhai-Macao Bridge – Red Edition; the flagship offering, Hearts of Loyalty – Party History on the Lingdingyang Sea; and the extension product, Pearl River Bay Tour – I Love My Country. Coinciding with the 100th anniversary of the Communist Party of China, the red educational tourism line successfully expanded in depth and reach, resulting in a significant surge in sales.

7.2. *Educational tourism brand renewal strategy*

As educational tourism policies continue to evolve and consumer demands shift, the content and presentation of educational tourism brands must also adapt and renew. Given the cyclical nature of product lifecycles—spanning development, introduction, growth, maturity, and decline—the essence and expression of a educational tourism brand cannot remain static. Particularly in today's market, where brand over-marketing is prevalent, it is essential to refine brand perception positioning. The key lies in employing simplified and straightforward messaging to help consumers understand how one brand stands apart from others.

At its core, the brand renewal strategy for educational tourisms should align with the development cycles of the target audience. It examines the consistency between consumer feedback and brand delivery across various age segments. For example, younger children are typically drawn to nature-themed or science-based tours, so greater emphasis should be placed on hands-on science and discovery experiences. With their idealism and enthusiasm for life, adolescents tend to gravitate towards exploratory and motivational educational tourism programmes that involve active participation and personal development. Older youth, who seek individuality and expression, prefer immersive and adventurous educational tourisms, such as nature expeditions, intangible cultural heritage experiences, sports training, and narrative-based role-play activities.

Therefore, the heart of the brand renewal strategy is to identify and focus on each product's core keyword or defining attribute. A brand can continuously reinforce consumer recognition by emphasising the most compelling selling point and maintaining a sharply defined category focus. Tactics include updating brand visuals and enhancing product offerings to ensure ongoing relevance and appeal.

7.3. *Educational tourism brand alliance strategy*

At its core, a brand alliance involves inter-organizational collaboration to unlock hidden value and form a larger, unified educational tourism brand network. Given the fierce market competition and serious product homogenisation, educational tourism providers must highlight their unique competitive strengths within each educational tourism value chain segment. Once a brand alliance is formed, participating enterprises can share distribution channels, integrate resources, reduce product launch costs, and amplify brand influence. For small to mid-sized educational tourism providers, the following two brand alliance strategies are efficient:

(1) **Cognitive Brand Strategy**
This approach involves multiple educational tourism enterprises adopting a unified brand identity for joint promotion and visibility. For instance, in May 2021, Sichuan Province launched the Geological Educational tourism Base Alliance, in which 45 certified bases combined forces to align with the local compulsory education course, "Lovely Sichuan". Integrating the curriculum and brand alliance helped boost educational tourism development across the province. Similarly, Leshan not only launched the Leyou Jiaxue educational tourism brand alliance, with standardised brand usage across logos, teaching tools, courses, and promotional materials, but also initiated a talent alliance to standardise the training of educational tourism professionals.

(2) **Complementary Brand Alliance Strategy**
Since every city possesses its distinctive educational tourism brand, a complementary brand alliance brings together neighbouring cities to create cooperative networks that optimise market synergy. For instance, Luoyang in Henan Province allied with Jiaozuo, Pingdingshan, Sanmenxia, and the Jiyuan Demonstration Zone to establish the Luoyang Metropolitan Educational tourism Alliance. This initiative aimed to develop a standardised, safety-focused, and regularly operated educational tourism cooperation mechanism across cities. Each city's brand concentrates on differentiated development, avoiding redundancy and emphasising local cultural attributes. With a market-driven approach, these cities creatively built interconnected educational tourism brand systems. They shared resources, cross-promoted, and regularly organised student exchange activities, while leveraging the educational influence of educational tourisms to promote coordinated regional economic and social growth.

8. Chapter Summary

This chapter primarily elaborates on the fundamental connotations of educational tourism branding, methods of brand value extraction, steps for competitive brand analysis, and the pathways to building and integrating a educational tourism brand into the audience consumption cycle. It also explored strategies for brand renewal. From a market perspective, the foundation of a sustainable educational tourism brand lies in the value of its products, its cultural attributes, and its unique identity.

9. Post-chapter Summary

(1) How can the educational tourism product branding strategy be considered equivalent to product development?
(2) Do most current educational tourism brands lean more toward brand strategy or advertising creativity? Provide examples and analysis.
(3) What are the methods for refining the core value of a educational tourism brand?

(4) What are the brand-building pathways for educational tourism products? Based on your region's current educational tourism landscape, write a short brand positioning report for a regional educational tourism initiative.
(5) Briefly describe the strategic approach to educational tourism product branding.

Chapter 7

Educational Tourism Market Segmentation and Target Market Strategies

Learning Objectives

1. Understand the categorisation and market supply of educational tourism products.
2. Understand the key points of educational tourism product design.
3. Understand the marketing strategies of educational tourism products.

Knowledge Framework:

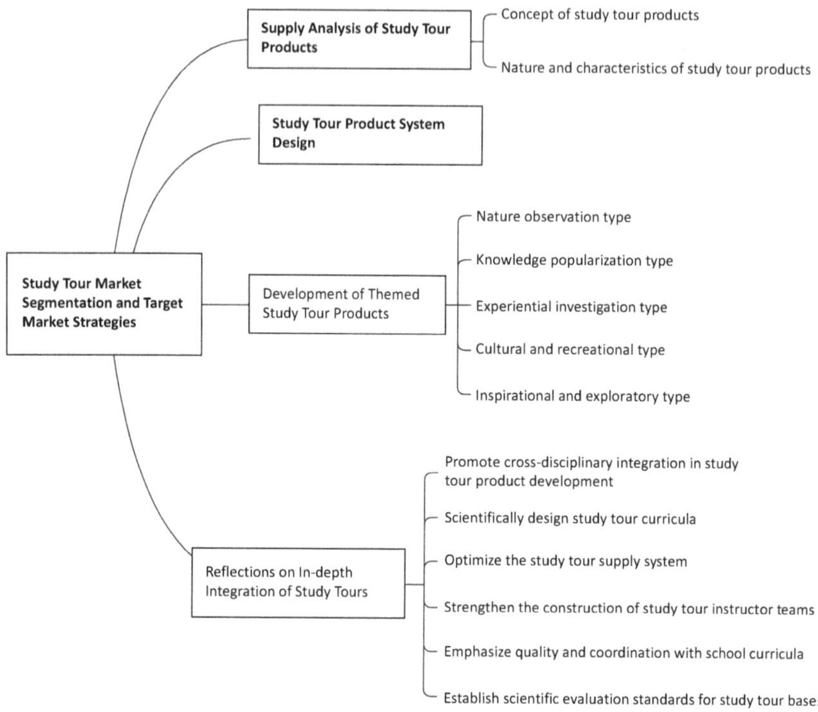

Learning Focus

1. Understand the conditions for designing educational tourism products.
2. Refine educational tourism products and understand the supply in the educational tourism market.
3. Learn the conditions for creating themed products.
4. Understand educational tourism products in different fields through classic case studies.

Contents

1. Case Study

Since the release of the Opinions on Promoting Educational tourisms for Primary and Secondary School Students by 11 government departments, including the Ministry of Education, in 2016, a series of education and tourism policies have supported the rapid growth of the educational tourism market. Integrating leisure agriculture with parent-child educational tourism demands has led to high-traffic projects combining rural sightseeing, cultural experiences, science education, and family learning.

I. *Rooted in Industry: Expanding Science Education Educational tourisms*
Miguoguo Town, situated in Shanjiahu Town, northeast of Zhuji City in Zhejiang Province, was established in 2013. By 2018, its total area had expanded to 3,152 mu, which includes 1,200 mu dedicated to dragon fruit plantations. In the subsequent years, Miguoguo Town transformed into an integrated park that combines planting and breeding, agri-product processing, leisure tourism, educational training, and creative industries through efficient land use and the integration of primary, secondary, and tertiary sectors.

Miguoguo Town features themed areas such as the Agricultural Science Museum, Happy Farm, Fruit Picking Zone, Student Field Laboratory, and Dragon Fruit Processing and Exhibition Zone. Visitors can engage in farming activities and rural life while observing agricultural production, highlighting the scientific educational value of educational tourism products. On weekends and holidays, the town experiences high tourist traffic; educational tourism groups fill the gap during weekdays, ensuring steady visitation even during the off-season. Moreover, introducing summer camps and winter sports programmes helps address the seasonal challenges of the hot climate in southern regions. In 2019, the town welcomed over 800,000 visitors.

II. *Based on Nature: Promoting Ecological Education*
Taiwan's Taiyi Ecological Education Farm, established in 1990, spans 0.23 square kilometres and is renowned for its extensive flower cultivation and picturesque beauty. Upholding the principle of "experiencing nature", the park connects visitors with nature's vitality and rural charm. With the design theme "Adding Fragrance to Life and Observing the Wonders of Growth from a Seed", the park provides hands-on activities and immersive environments, including a DIY pressed flower studio, cultural galleries, a floating garden restaurant, romantic flower bridges, the Temple of the Flower Spirit, the Root World, flower and vegetable cultivation zones, a seed-planting demonstration area, fantasy tea rooms, a butterfly pavilion, and an organic farming zone.

Visitors participate in flower pressing, observe plant growth, and engage in farming experiences that combine education with immersion in rural life.

The "Agriculture + Tourism + Education" model opens new possibilities for the parent-child educational tourism market. A robust industry foundation and distinctive products, capitalising on local industrial strengths, facilitate the development of competitive experiential courses. Meanwhile, education remains central to parent-child educational tourisms. Ecological resources should be meticulously crafted and refined into well-structured nature education programmes, establishing a solid foundation for future park development.

1.1. *Reflection questions*

1. What changes has the cross-sector integration of educational tourism products brought to local development?
2. How can educational tourism products be designed based on the characteristics of different industries?

2. Analysis of Educational Tourism Product Supply

In his essay Xing Zhi Xing, Chinese educator Tao Xingzhi wrote, "Action is the beginning of knowledge, and knowledge is the fulfilment of action".

As a form of travel, educational tourisms have a long history, evolving from ancient educational journeys and modern academic excursions. They embody and advance the traditional Chinese educational ideal and humanistic spirit of "reading ten thousand books and travelling ten thousand miles", becoming an indispensable component of contemporary holistic education.

Broadly defined, educational tourisms refer to specialised travel experiences that are research- and inquiry-based. These cater to a broad audience; any group that conducts exploratory visits to a educational tourism destination can achieve educational goals through travel, such as parent-child tours, summer camps, winter camps, themed educational tourisms, or corporate team-building. In a narrower sense, educational tourisms are student-centred, aiming to develop students' abilities by combining school-based learning with experiences beyond the classroom. Through experiential and open-ended education, they complement traditional methods. Students gain enjoyable travel experiences, broaden their horizons, and enhance their knowledge, making educational tourisms a meaningful life experience during youth development.

Since the concept of "educational tourism" was introduced, it has evolved into a diversified consumer structure. Schools, travel agencies, cultural institutions, or government-established educational tourism management bodies typically carry out implementation. Against a rapidly developing cultural tourism industry, educational tourism content is becoming more specialised, and the experiential aspect continues to grow. The "educational tourism +" model is emerging as a key strategy for developing educational tourism products.

However, with the rise in popularity and rapid expansion of educational tourisms, numerous issues have emerged, such as a lack of understanding of the definition, programmes focusing solely on sightseeing rather than learning, the absence of officially recognised standards for educational tourism instructors, uninspiring product design, and underdeveloped theoretical foundations.

Consequently, the actual value of educational tourisms lies in the depth of learning. At their core, educational tourisms emphasise the collaborative efforts of governments, schools, experts, and professional institutions to explore various educational tourism resources and develop distinctive, scientific, rich, and high-quality products. These should be categorised by theme to meet the needs of different age groups and effectively reflect the principle of "the unity of knowledge and action" within educational tourism education.

2.1. *Concept of educational tourism products*

Educational tourism products refer to the courses and services designed to meet the needs of educational tourism education in China. These are based on the characteristics of the destination, such as nature, culture, history, art, and science, emphasising inquiry-based learning and hands-on practical experience. The primary aim of these products is to achieve the integration of "tourism + education".

High-quality educational tourism products enhance the entire educational tourism process's educational, practical, and integrative nature. They help to establish a multidimensional, interconnected educational system that aligns closely with formal school education, thereby ensuring the widespread implementation of educational tourisms.

As the content of educational tourism education becomes more specialised, many diverse programmes are emerging, driving the continuous development and renewal of educational tourism products. "Cross-sector integration" has become a mainstream trend. Existing educational tourism bases are increasingly merging with outdoor training, aesthetic education, traditional culture, nature education, general practice, rural experiences, and patriotic education, forming a vital and comprehensive product system within the educational tourism market.

2.2. *Nature and characteristics of educational tourism products*

Since the Ministry of Education and ten other departments jointly issued the Opinions on Promoting Educational tourisms for Primary and Secondary School Students in December 2016, educational tourisms have rapidly become a popular integrated form of education. However, the quality of educational tourism products varies significantly. After years of experimentation, the consumption of educational tourisms has gradually matured, and market demand has evolved accordingly.

Educational tourism products are education-oriented at their core. They are not traditional tourism offerings but a new "study + travel" product, relying on cultural resources as a foundation, education as the means, and travel as the vehicle. Educational tourisms aim to cultivate skills, broaden horizons, and enhance aesthetic appreciation through experiential travel, emphasising the "study" component. Therefore, these products are developed around the destination's thematic ideas or academic value and bear the dual identity of educational and travel-based experiences.

Educational tourism products represent a comprehensive service system. Providing differentiated services that cater to diverse groups and their specific educational goals is key to ensuring quality. Each product reflects a distinct educational method. The completeness of essential elements—educational tourism bases, curricula, travel routes, instructors, and support services—determines whether a educational tourism can be successfully executed. The curriculum, transformed from resources, helps turn passive observation into active participation. Well-designed routes integrate various curriculum elements. Instructors play a pivotal role in guiding, explaining, and connecting key knowledge points, while support staff ensure the smooth delivery of basic services like accommodation, food, and transport. These components are interdependent and indispensable, making this new form of education irreplaceable. Therefore, educational tourism products must provide accurate and refined offerings tailored to different target groups.

(1) **Integration with School Curricula**
Educational tourism products should align with both schools and educational tourism institutions by complementing and enhancing school-based courses. Many destinations have a natural connection to school subjects, so products must work in tandem with curricula while adhering to sound educational principles. These products aid in delivering exploratory and innovative instruction, bridging in-school and out-of-school education.

(2) **Support Across Educational Stages**
As an extension of formal education, educational tourism products must cater to the specific learning goals of primary, middle, and high school levels. Drawing on local cultural resources, they should offer opportunities to "learn through travel" and "travel through learning." Since implementing the "Double Reduction" policy, parents have become more discerning, and schools have emphasised customising and planning educational tourisms. These programmes have become integral to well-rounded student development.

(3) **Products for Adult Learners**
With rising expectations for spiritual and cultural fulfilment, educational tourism products should also cater to adult learners. The demand has extended beyond schools into society at large—corporations now organise educational tourisms for team building, and elder colleges have introduced senior educational tourisms.

Individuals are also demonstrating a growing interest in adult-oriented educational tourisms. These products necessitate a higher standard of content and structure, as adult learners seek personal growth and a sense of identity through such experiences. To resonate with this audience, educational tourism programmes must meet elevated expectations regarding thematic design, professionalism, academic content, and service quality.

3. Designing Educational Tourism Product Systems

Scenic sites serve as a significant foundation and destination for educational tourisms, with numerous tours conducted through these venues. In recent years, the increasing share of educational investment and the shift in educational philosophies have encouraged integrating experiential education with tourism, rendering educational tourisms a market hotspot. This trend imposes new demands on the development of scenic sites:

Firstly, the overall environment of tourist attractions should be optimised by transforming appealing resources, industries, and various elements into new types of tourism products. Secondly, comprehensive tourism services should be developed to realise integrated tourism industry connections across regions. Establishing clusters of distinctive tourism products can promote the integrated development of diverse offerings, formats, and industries, leveraging the synergy of "Tourism+" to incorporate multiple sectors and create a fully integrated industrial chain and tourism ecosystem. Lastly, mutual growth between tourism and other sectors should be encouraged to establish a large-scale, interconnected development framework. Scenic areas across the country have strengthened collaboration with travel agencies, training centres, educational tourism institutions, and schools, yielding effective results. Furthermore, introducing educational tourism projects has positively impacted scenic sites by increasing popularity, diversifying offerings, and extending the value chain.

To establish a robust and influential educational tourism brand, scenic destinations must adapt to market demands by continuously integrating resources from universities, research institutions, and educational tourism bases. Products and routes should be customised to meet the diverse needs of various educational tourism groups. Furthermore, regional resource integration and multi-site coordination should be strengthened to create a shared network of tourist traffic and educational tourism resources. Proactive regional collaboration must be encouraged to foster the development of educational tourisms across all regions.

But how should scenic sites organise educational tourism activities? How can they design educational tourism curricula based on their unique characteristics to meet market demands? To address these questions, picturesque sites in China must undergo a thorough transformation—one that restructures services to align with the requirements of educational tourisms.

First, scenic sites should integrate educational tourism resources. As previously discussed, this involves aligning surrounding resources with market demand, identifying unique educational attractions, and customising distinctive educational tourism products and value chains tailored to the site's characteristics.

Secondly, scenic sites must align with school education plans. According to national and local policies and the Ministry of Education's requirements, educational tourism products should be tailored for students at various educational stages. Educational experts, psychologists, and tourism planners should collaboratively develop educational tourism offerings for scenic sites. Most importantly, sites must segment their markets and audiences to create targeted curriculum systems. For instance, by analysing young people's cognitive levels and needs, in conjunction with the curriculum of schools and the site's available tourism resources, scenic destinations can design educational tourism programmes in areas such as nature, geography, science, humanities, and experiential learning. These should be created with an emphasis on participation, engagement, and interaction to address diverse needs.

Thirdly, scenic destinations must enhance facilities and elevate service standards to foster the development of educational tourisms. Infrastructure should meet the requirements of various educational tourism bases and camps at multiple levels, including transportation, attractions, support services, environmental upgrades, and reception facilities.

Finally, smart educational tourisms should be implemented through a "Educational tourism + Big Data" platform. Scenic sites can utilise internet technology to build customer databases and use big data to effectively link landscape resources, educational tourism participants, and staff. This enables real-time data sharing, efficient control, and early warnings across infrastructure, logistics, tour services, and safety management. This approach improves scenic site management and allows parents to monitor safety, enhancing their trust in the site.

This chapter will analyse classic case studies of scenic site educational tourisms to illustrate how effective educational tourism product systems can be designed and implemented.

3.1. *Case introduction*

Educational Tourism Case: The Chocolate Factory

Transforming industrial factories into educational tourism resources is a classic approach to unlocking the educational tourism market. In Taiwan, sightseeing factories have evolved into cultural and creative industries, building on cultural content as the foundation and core, constantly launching trend-aligned products. These combine industrial production experiences, parent-child educational tours, innovative formats (e.g. themed museums, restaurants, parks), and speciality shopping to generate new economic vitality.

In the 1990s, Taiwan's industrial development was relatively slow, but some industrial brands recognised market trends early and took the lead in transforming into sightseeing factory projects. These sightseeing factories have become popular for family and educational tourism activities in Taiwan.

1. **Taoyuan Bade Chocolate Republic**

   Situated in Bade, Taoyuan, Chocolate Republic is the first chocolate museum in Southeast Asia. The design of the venue and park subtly integrates chocolate elements throughout.

   In recent years, Chocolate Republic has focused on developing parent-child programmes. From comprehensive demonstrations of chocolate production, packaging, and creation to interactive chocolate DIY sessions for families, these experiences offer educational and entertaining outcomes while enhancing parent-child interaction, receiving widespread acclaim. Chocolate Republic has attracted numerous visitors by transforming its established manufacturing technologies into educational tourism resources and introducing chocolate-themed courses and practical experiences. It also provides an extensive range of themed products, such as speciality chocolate snacks and beverages, to cater to diverse consumer needs.

Chocolate Republic features three main areas: the first floor serves as a commercial zone, the second floor is dedicated to chocolate knowledge, and the third floor, the most popular among children, is the DIY experience zone. It is both interactive and educational.

The exhibition showcases over 3,500 years of chocolate history and provides detailed information on cocoa-related agriculture, illustrating how cocoa beans are transformed into delicious chocolate. Visitors can explore independently or join guided tours for engaging learning experiences. Additionally, professionally-led DIY chocolate-making classes enable participants to create unique chocolates.

The factory tour walkway showcases production lines of the brand's famous products, such as the 77 Nougat Bar, Big Boro, and Puff Pastry Line, providing visitors a front-row seat to the manufacturing process. A visit to Chocolate Republic is akin to reading a richly illustrated chocolate book or a professional textbook on the baking industry.

Signature Products and Fun Activities: Hands-On Chocolate Making: (1) Under the guidance of instructors, participants create chocolate bars and treats (blending unique Taiwanese flavours with cocoa), embellish their creations, and enjoy the experience with the assistance of friendly pastry chefs. (2) Seasonal Events with a Twist: Activities such as "2019 Chocolate Fortune-Grabbing Ceremony" and "Let's Go Party!" offer engaging and festive experiences.

2. **Sophisca—The Candy Wonderland**

   As Taiwan's first enterprise dedicated to candy production, Sophisca established its candy factory in 2012 to integrate retail and DIY experiences in response to local market demands. In addition to offering a wide range of aesthetically appealing confectionery products, the Sophisca Candy Factory has also launched engaging parent-child tours. The factory has transformed the traditional visit into an interactive and enjoyable experience by showcasing the candy-making process and encouraging hands-on family participation. Guests leave not only with sweet treats but also with treasured memories.

   By transitioning into a cultural and creative industry, Sophisca's sightseeing factory enhances children's social skills and ecological awareness through candy craft exhibitions and interactive learning.

Its flagship factories in Hsinchu and Yilan sell a variety of colourful, uniquely designed sweets and boast DIY learning as a significant highlight.

DIY Workshops: The parent-child DIY classroom offers chocolate-making sessions, where each family can enjoy the satisfaction of creating their own treats. With expert guidance, everyone can craft their unique chocolate delight—delicious and enjoyable.

Live Demonstrations: Interactive experiences make DIY sessions exceedingly popular, often filled to capacity. The factory features transparent production rooms where staff showcase candy-making in real time to reduce wait times. Guests are even treated to freshly made sweets straight from the production line, enhancing the immersive experience.

Ecological Education: Beyond the factory, the pond filled with tadpoles captures children's curiosity. This natural setting serves as an outdoor environmental classroom, providing parents with a perfect opportunity to teach their children about life and the environment. Families can also enjoy observing fireflies together at night, offering an enchanting way to learn about environmental conservation.

Chocolate DIY: Children can express their creativity using moulds and colourful edible markers to craft delightful, edible art. With parental guidance, they produce vibrant and joyous chocolate pieces. They also have the opportunity to embellish their creations with hard sweets and marshmallows, adding bursts of colour and texture to enhance the appearance and flavour of their chocolate.

1. **Product Design**
   A well-developed experiential factory should feature a diverse product system, a comprehensive service and operations framework, and solid profit-generating capabilities. Creative product design is essential for unlocking business potential—it must be enjoyable and innovative, emphasise uniqueness, and create emotional resonance with visitors. Furthermore, professionalism is crucial. Successful design often involves collaboration between seasoned external teams and in-house specialists.

2. **Key Development Points**
   Industrial educational tourisms typically focus on experiential factories and are further enhanced by the potential of cultural and creative

industries. These destinations aim to engage the five senses and utilise localised packaging to offer interactive and educational experiences. A strong emotional and cultural core, along with thoughtful interactive packaging, are essential components.

i. **Established Industrial Base**
Most experiential factories pride themselves on being “firsts” in their field, such as Taiwan’s first microbiology-themed museum or the first to create a signature scallion egg roll. A strong and reliable industrial foundation and standout features form the backbone of a successful experiential factory.

ii. **Unique Product Offerings**
From a product development perspective, the structure generally adheres to a “core product + derivative products + supporting services” model. The core product is the factory’s original item, whilst derivative products introduce creative extensions, such as ginseng-flavoured ice cream or handmade soy pudding from a soy milk brand. Guided tours are a common and essential service, and food-related factories often offer on-site dining to enhance the overall experience.

iii. **Creative Cultural Derivatives**
Experiential factories in Taiwan often emerge from the perspective of building and promoting local brands. By weaving engaging brand narratives, they showcase the beauty of local people and culture, communicate corporate values, and highlight Taiwan’s rich heritage. This storytelling approach enhances product value and has become a mainstream marketing strategy. For instance, creative combinations like LOTON organic soy milk, scallion egg rolls, lightbulb bubble tea, and ginseng ice cream—paired with innovative production techniques—serve as major attractions for visitors.

iv. **Interactive Experiences that Energise Industrial Tourism**
Experiential factories are not merely showrooms for products; they serve as stages for showcasing an entire industry. These venues aim to captivate visitors through multisensory engagement, blending elements such as cultural reinvention, concepts from the experiential economy, and art exhibitions with enjoyable, educational activities. The ultimate goal is to reshape consumer behaviour through immersive experiences, “performance spaces”.

Among interactive offerings, DIY workshops are the most common and effective. These activities—such as making products, participating in fun classes, or engaging in hands-on crafts—bring education and entertainment together. DIY also serves as a powerful form of product marketing, offering visitors an intimate encounter with industry culture. Popular options include DIY wine blending, product packaging, cookie making, candle crafting, and more. Combined with games, educational sessions, and tastings, they form a robust and memorable experience ecosystem.

4. Developing Themed Educational Tourism Products

According to the *Educational tourism Service Specification* (LB/T 054—2016) issued by the Ministry of Culture and Tourism, educational tourism products are classified into five categories based on resource type: nature appreciation, scientific knowledge, hands-on investigation, cultural leisure, and motivational development. This section explains how to develop themed products through classic case examples.

4.1. *Nature appreciation*

Nature appreciation educational tourism products focus on national-level attractions such as World Natural Heritage sites, mountains, rivers, lakes, oceans, grasslands, and deserts—these resources are among the most popular. By analysing and leveraging a site's unique natural and cultural elements and effectively translating them into educational content, such products encourage cultural exploration and help participants develop an appreciation for nature and a consciousness for environmental protection.

Model Case 1

Zixuetang "Exploring the Panda of the Plant World"

1. **Target Participants**
   All students from Zixuetang's Advanced Class.

2. **Program Objective**
   To explore the propagation and growth conditions of the dove tree (Davidia involucrata) in the Longcanggou region.

3. **Key Features**
   On-site scientific field research.

4. **Course Content**
   Over the course of three days, students carry out comprehensive fieldwork in the Longcanggou region to examine the environmental factors influencing the growth and reproduction of the dove tree. Observations take place at three altitudes—600 m, 1000 m, and 1500 m—focusing on temperature, humidity, topography, soil and water pH, and associated plant species. Students also analyse the structural properties of dove tree seeds, including shell hardness, to better understand the challenges related to reproduction.

5. **Course Wrap-up**
   Students form groups to select specific research focuses and compile their findings into a scientific mini-paper as a final summary of the educational tourism.

Model Case 2

Zixuetang "Fairy Tale in Dong Village"

1. **Target Participants**
   All students from Zixuetang's Foundation Class.

2. **Program Objective**
   To gain a thorough understanding of the customs and lifestyle of the Dong ethnic group in Chengyang, Sanjiang.

3. **Key Features**
   A combined curriculum focused on exploring folk culture, experiencing intangible cultural heritage (ICH), and encouraging children's sketching and painting.

4. **Course Content**
   Students immerse themselves in Dong culture by living with local families in Chengyang Sanjiang for six days, in groups of four. During this time, they experience the daily life of the Dong people—food, clothing, housing, and customs. Students engage in traditional activities such as tea seed harvesting, oil pressing, spinning, weaving,

batik, tailoring, and making sticky rice cakes. A highlight of the course is learning and performing the traditional Grand Dong Song, which offers a fully immersive cultural experience.

5. **Course Wrap-up**
   On Day 7, students present their sketches of iconic Dong architecture, such as the Wind and Rain Bridge, and handcrafted batik works in the local drawing base's conference hall. The programme concludes with a Hundred Family Banquet, where students perform the Grand Dong Song.

4.2. *Science and knowledge-oriented educational tourism programs*

Science and knowledge-oriented educational tourism products mainly include resources such as museums, science and technology centres, themed exhibitions, zoos, botanical gardens, historical and cultural heritage sites, industrial projects, and research facilities. These resources are particularly popular among educational tourism groups focused on holistic education.

Model Case 1

Exploring the Hexi Corridor and Appreciating the Grottoes of Gansu

Gansu is known as the "Land of Grottoes". Along the 900-kilometre scenic stretch of the Hexi Corridor, one can find snow-capped mountains, glaciers, deserts, lakes, plateaus, and colourful hills—spectacular sights interwoven with the rich heritage of the Silk Road. This region, inhabited by diverse ethnic groups for centuries, presents an ideal destination for cultural and educational exploration.

1. **Target Audience**
   Adults

2. **Program Objective**
   Travel along the Silk Road to gain an in-depth understanding of its natural landscapes and cultural landmarks, and foster a strong sense of cultural identity and pride.

3. **Curriculum Overview**
   2017, the Dunhuang Academy has officially taken over the management of six major grotto sites: the Maijishan Grottoes Art Research Institute, the Bingling Temple Cultural Heritage Institute, the Beishiku Temple Preservation Institute, as well as the Yulin Grottoes and the Western Thousand Buddha Caves. By leveraging its academic and tourism expertise, the Dunhuang Academy's Mogao Classroom developed the signature educational tourism programme "Exploring the Hexi Corridor and Appreciating the Grottoes of Gansu". This immersive programme features a historical overview from the Sixteen Kingdoms and Northern Dynasties through the Tang, Song, Western Xia, Yuan, and Ming dynasties, focusing on grotto sculptures and mural art.

4. **Program Highlights**
   i. **Expert-Led Learning:** Specialists and senior guides from the Dunhuang Academy comprise the teaching team, delivering insightful lessons on the cultural heritage of the Silk Road.
   ii. **Exclusive Access to Grottoes:** Carefully curated routes combine premium and rarely accessible caves, providing participants with a guided, immersive experience of Gansu's stunning grotto art.
   iii. **Creative Workshops:**
      a. **Mogao Dream:** Clay Painting Experience: Participants replicate scenes from Dunhuang murals on clay tablets, employing traditional techniques alongside their personal interpretations.
      b. **Discover Dunhuang Colours:** Mineral Pigment Production: Uncover the secret to the lasting brilliance of Dunhuang murals by creating pigments from natural minerals.
   iv. **Full-Service Support:** The elite team at the Mogao Classroom provides comprehensive support throughout the programme, with experienced instructors ensuring a seamless and enriching journey.

In today's era of advanced technology, educational tourisms centred around science and innovation have also emerged. These programmes often involve interactive learning through VR, AR, and 3D/4D technology. Destinations typically include science museums, research centres, and tech parks, allowing participants to engage directly with cutting-edge developments in a hands-on environment.

Model Case 2

Exploring the Frontier of Photovoltaics: Validating Truth Through Practice

1. **Program Background**
   The Dunhuang Photovoltaic Industrial Park is a comprehensive renewable energy hub. Anchored in solar power generation, it encompasses the manufacturing of solar panels, inverters, wind turbines, and high-capacity lithium-ion storage batteries. Its vision includes promoting clean energy applications and establishing a national centre for solar power testing and training. It is China's first government-approved "one-million-kilowatt-level solar PV demonstration base". In 2016, the park was rated as a national 3A-level tourist attraction. As Dunhuang integrates new energy industries with tourism, the development of science-themed educational tourisms has flourished. The New Energy Exhibition Hall features themed areas such as the introduction hall, science education zone, multimedia sandbox, and planning area. With over 30 panels, more than 40 models and exhibits, and advanced technologies including virtual greeters, light and sound sandboxes, and nano-touch screens, it comprehensively showcases the history of energy, the status and future of renewables like wind, solar PV, and solar thermal in China, as well as the development of Dunhuang as a renewable energy city. It also houses the 10MW molten salt tower CSP plant, Asia's first and the world's third solar-thermal plant capable of 24-hour continuous power generat.

2. **Course Objectives**

   i. Enable students to gain insight and inner enrichment through the integration of theoretical knowledge and hands-on exploration in science and geography.
   ii. Use simple language, vivid analogies, and clear experiments to demystify complex scientific principles, helping students form intuitive connections with science and cultivating curiosity.
   iii. Promote scientific thinking, foster creativity, and encourage a passion for research through inquiry and exploration.
   iv. Develop students' hands-on skills, innovation capacity, and teamwork abilities, laying a strong foundation for holistic education in higher learning.

3. **Related Disciplines**
   Morality & Law, Physics, Geography, Chinese Language, Mathematics.

4. **Pre-Class Questions**
   i. What's your current understanding of photovoltaic energy?
   ii. What standard devices or services in daily life are powered by photovoltaic technologies?
   iii. How do you view solar power as a sustainable energy solution?

5. **Curriculum and Structure**
   This experiential course immerses students in hands-on scientific exploration, bringing STEM to life beyond the classroom. The program is divided into four modules:

   i. Science Knowledge Primer
   ii. Guided Tour of the Industrial Park
   iii. Solar-Powered Car Speed Race
   iv. Construction of a Model Solar Thermal Power Station

   By incorporating theory, observation, and hands-on practice, this program offers a dynamic, memorable learning experience that deepens students' understanding and enthusiasm for science. The course is structured into four main formats:

   i. **Observation Tour**
   Students visit a comprehensive exhibition hall focused on photovoltaic knowledge and observe large-scale solar thermal and photovoltaic power generation projects in action.
   ii. **Field Investigation**
   Students explore how solar photovoltaic systems replace conventional power generation in everyday life and industrial production.
   iii. **Hands-On Experience**
   Students simulate the working principles of solar power and enjoy assembling and operating a solar-powered device.
   iv. **Scientific Inquiry**
   Students investigate the technological foundations behind solar energy as a form of renewable energy.

6. **Activity Rules**
   i. Teachers assess whether students complete the solar car assembly within the given time. A complete build must include: chassis,

steering wheel, solar panel, battery, and car frame, and must be able to move using solar power.

ii. Teachers also evaluate the design of the car for practicality and aesthetics.

iii. After assembly, each group selects a representative to participate in a race.

iv. The group with the fastest completion time wins and receives a commemorative prize.

7. **Study Focus**
   i. Advantages and disadvantages of solar energy.
   ii. Methods of utilising solar power.
   iii. The transformation of energy from one form to another.

8. **Learning Challenge**
   i. Understanding and following the assembly manual.
   ii. Successfully building a functional solar-powered vehicle.

9. **Post-Class Extension**
   i. What other emerging solar-powered devices or installations do you know?
   ii. Do you understand how they work?
   iii. What other new energy sources are being used today, and what are their functions?

Huijing Xincheng Experimental Primary School "Sea-Land-Air Trifecta" Technology-Themed Winter Camp

1. **Course Objectives**
   This winter camp aims to broaden students' horizons and ignite their curiosity about the world. It focuses on cultivating four essential abilities: verbal communication, logical thinking, emergency response, and emotional regulation. In addition, it nurtures three key values: a love for culture, a healthy mindset, and a strong sense of national identity.

2. **Course Highlights**
   (i) Sea: A closer look at military shipbuilding—unveiling the might of national defence
   Students first visit the Huangpu Wenchong Shipyard Historical Museum to learn about the century-old legacy of China's

military shipbuilding industry and the perseverance of shipbuilders through generations. They explore the structures, types, and functions of various vessels. Next, they get up close with naval ships docked at the port, gaining a deeper understanding of active-duty warships and the strength of China's maritime technology, instilling a sense of patriotism and national pride. The program also includes a visit to the historic Ke Bai Dry Dock (now part of the Guangzhou Huangpu Shipyard), which is not only the first granite dry dock in the Far East but also China's first modern stone dock. Built in 1851, it is a significant national industrial heritage site that bears witness to China's modern struggles and triumphs. Finally, students work together to assemble a model of the "Guangzhou 168" missile destroyer and engage in hands-on learning about Morse code.

(ii) Land: Exploring the science and technology behind urban rail transit

Through a visit to the Guangzhou Metro Museum, students gain insights into the development and cultural significance of urban rail systems while completing guided learning activities. They then explore the workings of tunnel boring machines, examining the only full-scale model of its kind in the country, to understand how subway tunnels are excavated and how internal systems operate. In addition, students engage in geological discovery by observing a realistic recreation of karst landscapes and geological layers, enabling them to touch and study Guangzhou's complex underground structure, thus gaining intuitive knowledge of geology and hydrology.

(iii) Air: Understanding the evolution of China's aerospace industry

Students attend an introductory aviation course where they learn about fundamental flight concepts under the guidance of a professional pilot, developing a multidimensional understanding of aviation systems and classifications. They observe airport zones, identify common aircraft types, and witness the take-off and landing of planes at close range. Students also visit themed spaces such as the "Time Tunnel", the "Sky-Sea Corridor" that highlights Silk Road culture, and the Lingnan cultural graffiti wall, prompting reflection on cultural confidence and the transmission and innovation of traditional heritage. Transforming into "Junior Chinese Pilots", they use AR technology and a flight simulator to experience piloting in a virtual cockpit and gain a hands-on understanding of a pilot's role. To explore the mysteries of the universe and China's aerospace achievements, students ride the "Huaxia Ark" and wear panoramic VR headsets to simulate a moon landing, experiencing what it feels like to be an astronaut.

4.3. *Experiential and exploratory programs*

Experiential exploratory educational tourism programs mainly draw upon resources such as farms, internship bases, summer camps, and team-building training centres. The American Camp Association defined camp-based education 1988 as "a sustained experience in an outdoor setting, characterised by group living, that is creative, recreational, and educational in purpose". According to a representative from Cedar Camp, camp-based education is not merely about teaching content—it is about guiding children to develop a lifestyle. It is during moments of relaxation, away from competition, and in an environment where they feel free, that children grow most rapidly. These are when they have the best experiences, can reflect, and effectively communicate and collaborate with others.

Model Case

"Rivers, Oceans and Lenses": Exploring the World through Images
The "Rivers and Oceans Youth Film Academy" is a specialised educational tourism institution founded by media professionals and documentary filmmakers. Since its inception, the academy's flagship programme—the Rivers and Oceans Visual Storytelling Camp—has established a distinctive model: employing real-world environments as classrooms, film techniques as learning tools, and creative output as its primary goal. Students visit historical locations, immerse themselves in cultural settings, explore heritage sites, and produce original works. Documentary filmmaking is an ideal medium for exploration, documentation, and storytelling.

Visual storytelling demands a wide range of technical and creative skills. Students are required to learn photography, video editing, scripting, production, and media dissemination, while also developing abilities in interviewing, writing, art, music, critical thinking, emotional awareness, aesthetic judgement, and cultural insight. These competencies enable them to create original works and bring their creative visions to life.

The programme is typically divided into three units. The first unit concentrates on foundational courses and practical experience. Students acquire essential filmmaking skills such as directing, cinematography, aerial drone photography, scriptwriting, and editing.

The second unit involves on-location filming. Students form production teams of two to five members based on selected topics. Each team member takes on roles such as director, cinematographer, editor, or scriptwriter, collaborating to shoot and edit their individual or group projects.

The third unit focuses on screening and sharing. At the end of the programme, a showcase of student films takes place. Residents, filming participants, and peer volunteers are invited to watch and provide feedback alongside the students and teachers. These screenings allow students to give back to the communities they filmed. Once the programme concludes, instructors continue supporting students in sharing their work online—on platforms like Bilibili, iQIYI, Youku, and Tencent Video—and encourage participation in domestic and international film festivals to gain broader exposure and feedback.

Moreover, the programme integrates public welfare by offering scholarships to local youth volunteers. These volunteers serve as cultural ambassadors for their communities, guiding visitors through their hometowns, translating, learning filming techniques alongside other students, contributing collaboratively to the creative process.

Students produced nearly a hundred documentary short films in their first three years of implementation. These were screened on-site and at their home schools, and the works were showcased at significant events such as the Silk Road International Youth Photography Exhibition and the World Nomad Short Film Festival, earning multiple awards and recognitions.

Bella Town—A Large-Scale Ecological Parent-Child Town for Youth Quality Education

Bella Town is situated in Changsha, Hunan. It focuses on developing educational tourism products that combine non-powered amusement parks with educational courses. Centreing on quality education for youth, it integrates multiple formats including IP concepts, non-powered physical training education, life skills education, social awareness education, experiential education, nature education, art education, and leisure vacation, forming a large-scale ecological parent-child town for youth education.

Within Bella Town are nine themed non-powered amusement and training zones: *Flight Exploration*, *Wisdom Valley*, *Courage Slide*, *Climbing Relay*, *Full Acceleration*, *Forest Crossing*, *Strongest Brain*, *Fish Leaping to the Sky*, and *Modular Space*. Covering approximately 30,000 square metres, the park has more than 70 advanced non-powered facilities with training value for youth.

Through various playful and training activities, the park seeks to cultivate logical thinking and the courage to face challenges in young people while also enhancing their communication, organisational, and teamwork skills. The result is a large-scale, quality education park for youth, featuring a clear theme that merges education with entertainment.[1]

Cedar Camp—"Education is life, and life is education"

Cedar Camp (CEDAR) is a pioneering brand in camp education in China. Established in 2008, it was the first to propose the "Migratory Birds Camp" concept and developed an independent Migratory Birds Camp (MBC) education management system. Cedar Camp aims to dismantle the barriers between family, school, society, and nature, reconstructing a fifth-dimensional space centred on children, with the joint participation of parents and social forces. Its primary offerings include the original Migratory Weekend, Migratory Classroom, and Migratory School (local or cross-border), enabling young people to participate throughout the year.

Following the introduction of the "Double Reduction" policy, Cedar Camp has also been exploring the integration of educational tourisms and camp education. Both educational tourisms and camp education serve young people; however, educational tourisms are curriculum-based, whereas camp education is life-oriented. Educational tourisms are divided by age groups, while camp education can involve mixed-age groups. The two approaches can operate independently or include mixed-age projects, fostering a leap in thinking.

[1] *Source*: Official website of Bella Town.

Therefore, Cedar Camp is experimenting with a mixed-age educational model that combines "educational tourism + camp education," creating a more relaxed study environment that facilitates inter-age social interaction. In essence, educational tourisms and camp education share the same philosophy: "education is life, and life is education". When education becomes integrated into daily life and evolves into a lasting culture, the growth of young people becomes all the more promising.

4.4. *Culture and recreation-oriented*

Culture and recreation-oriented educational tourism products often appear alongside other types of programs. They mainly include resources such as various theme parks and entertainment film cities.

Model Case

"Encore Dunhuang"—A New Cultural Tourism Landmark for Gansu

Encore Dunhuang is the first immersive indoor theatrical experience in Northwest China. The production adopts a "streamlined" spatial viewing format, guided by a "time-travel" concept. It features six iconic historical figures and five representative scenes from Dunhuang's rich past, weaving them into the main storyline. By transcending time and space, the performance brings history closer to the audience, guiding them through a walking experience that reveals a thousand years of Dunhuang's stories.

The maze-like theatre is equipped with rare, large-scale mechanical stage systems, 3D projection technology, and mobile giant screens. Many of the special effects used are industry firsts, immersing the audience in a grand and fantastical visual realm.

Encore Dunhuang was developed with a strong commitment to historical accuracy and academic integrity. It serves as a cultural introduction for visitors, offering a deeper understanding of Dunhuang's historical significance. Here, the audience is not merely watching a show—they become part of the performance, experiencing a dynamic blend of narrative and environment in an unforgettable, immersive encounter.

a. **Producer:** Gansu Siku Cultural Development Group Co., Ltd.
b. **Director:** Wang Chaoge

c. **Project Goal:** To artistically present the spirit of the Silk Road—peace and cooperation, openness and inclusiveness, mutual learning, and shared benefit—as embodied in the Belt and Road Initiative. The production highlights Dunhuang's cultural role within this national strategy and aims to share the enduring legacy of Chinese civilisation with the world.

Midea Luhu Tea Culture Educational tourism Base

The Midea Luhu Forest Resort is Midea Real Estate's first travel-and-residence town with distinct Chinese characteristics. Spanning an area of 6,800 mu, it was constructed with a total investment of 30 billion yuan by Midea Real Estate, adhering to the standards of a national 5A-level scenic spot. It is a comprehensive town that combines "investment and wellness", integrating 15 major high-end resort facilities. Within the area lies Chuan Tea Plantation, which has a history of over 70 years. Its predecessor was the state-owned Gaoming County Mechanical Tea Demonstration Farm, established in 1951, and previously one of the most productive tea plantations in the Pearl River Delta.

Midea Real Estate has preserved the original structure of the tea plantation and transformed it into a cultural tourism town themed around tea culture, revitalising the Chuan Tea Plantation. Furthermore, the city features 800 student accommodation units constructed to five-star standards, capable of housing nearly 2,000 students.

The Midea · Luhu Tea Culture Educational tourism Base primarily hosts school educational tourisms, winter and summer camps, parent-child tours, and outdoor development activities. It can accommodate up to 1,000 students each day. Currently, it offers a variety of educational tourism courses, including the Foshan Midea Luhu Tea Culture Study Programme, Foshan Midea Luhu Exploration Kingdom, and Labour Education module.

Integrating educational tourism education into real estate projects serves both commercial needs and educational purposes. It facilitates the development of parent-child learning experiences rooted in real-world environments. Compared to traditional exam-oriented "rote learning", this open and multi-scenario study system is more vivid and engaging, aiding children in retaining knowledge more effectively—something parents also welcome.

Real estate developers are encouraged to create related cultural tourism products and, from the perspective of lightweight innovation, to transform their projects into part of the educational tourism industry chain. This approach explores the potential added value of the educational attributes within cultural tourism resources.

Integrating educational tourisms with real estate projects has enabled large-scale development featuring comprehensive facilities. However, it is essential to return to the core goals of experiential education. Building on existing base resources, the focus should shift towards developing well-structured, thematically distinct, and educationally meaningful camp education programmes. In particular, these should align with local policies regarding base admission standards, exit mechanisms, and evaluation systems, while also enhancing the supporting project conditions.

4.5. *Inspirational and development-oriented*

Inspirational and development-oriented educational tourism programs typically focus on skill-building activities, which are generally divided into three types: water-based, overseas, and on-site. These programs fully use resources such as revolutionary education bases, university campuses, national defence education centres, and military camps.

Model Case

Shaoshan Mao Zedong Memorial Hall — Red-themed Experiential Education

On 22 November 2017, the Ministry of Education officially announced the results of the first national evaluation of educational tourisms and practical education programmes for primary and secondary school students. Among them, the Shaoshan Mao Zedong Memorial Hall, recommended by the National Cultural Heritage Administration, was the only practical education base in Hunan Province selected as part of the first group of national-level educational tourism bases.

Shaoshan, the hometown of Comrade Mao Zedong, enjoys unique advantages in developing educational tourism programs due to its rich historical and educational resources. In 2018, Xiao Guoliang, then deputy inspector of the Shaoshan Administration Bureau, noted that the Shaoshan

Mao Zedong Memorial Hall is the only museum in the country dedicated explicitly to systematically presenting Mao Zedong's life achievements, thoughts, and personal character. It is also recognised as a national model base for patriotic education, a national base for clean government education, a national first-class museum, and one of the first demonstration units for Lei Feng volunteer service in public cultural facilities. The museum offers a wide range of patriotic education resources, with a complete and structured system that includes more than ten permanent and special exhibitions.

Since 2018, the Memorial Hall has treated educational tourisms as a key platform for delivering ideological education, patriotic values, revolutionary traditions, national awareness, and practical subject learning for primary and secondary school students. Drawing on its distinctive red education resources and considering the psychological development, learning abilities, and practical needs of students, the Hall has developed a variety of themed educational activities, including inspirational character-building experiences, immersive re-enactments, and ceremonies aimed at fostering aspirations and life goals. These educational tourism programs are diverse in format, clear in theme, and rich in features, covering history, culture, and experiential learning.

The Memorial Hall uses several important red education sites, including Mao Zedong's Former Residence, the Nan'an Private School, Mao Zedong Square, and the Mao Ancestral Hall. Anchored in Mao Zedong's cultural legacy, it has created a red-themed educational tourism route that expands and enriches the Memorial Hall's offerings. At the same time, in line with the curriculum characteristics and educational objectives of primary, junior secondary, and senior secondary levels, the Hall has developed tailored materials for each age group. For example, in the inspirational character-building program, primary school students are guided to create a love for reading through the lesson "Learning from Chairman Mao: Loving to Read, Reading Good Books". Junior secondary students are encouraged to develop unity and mutual support through the lesson "Learning from Chairman Mao: Unity Is Strength". Senior secondary students are encouraged to think about life choices through the lesson "Learning from Chairman Mao: Choosing the Right Path in Life".

While touring Shaoshan's core sites, students can walk along Mao Zedong Square and lay flower baskets at the foot of Mao's statue. They can visit the Mao Zedong Memorial Hall and his Former Residence, where they can learn about his life through historical artefacts and photographs. They can also take part in study ceremonies held at the Nan'an Private School, where the setting helps them remember Mao's virtues—his thirst for knowledge, his spirit of exploration, his respect for teachers, and his thoughtful and action-oriented character.

But how should red-themed educational tourisms be designed to create genuinely empathetic student experiences? From a practical standpoint, several aspects need to be addressed. Many existing red-themed programmes emphasise surface-level experiences but fail to engage students emotionally. To address this, historical scenes, supported by virtual and intelligent technologies, can be restored to simulate past experiences. From a disciplinary perspective, educational tourisms should facilitate students' understanding and internalisation of the values and significance of red culture through various academic lenses, enriching the content and depth of each programme. These tours should also nurture students' affection for the Party, the country, and the socialist system, instilling a sense of national pride and a deeper patriotic sentiment while guiding them towards healthy emotional development and value formation. Red-themed educational tourisms ought to integrate emotional education, ensuring they carry substantial cultural and spiritual significance that aids students in gaining a deeper understanding of the Party's illustrious history. At the same time, collaboration with revolutionary heritage institutions and broader social forces should be strengthened to jointly develop meaningful and impactful educational tourism programmes.

Moreover, improving the quality of instructors for red-themed educational tourisms is essential. These instructors should possess a solid understanding of revolutionary history and culture and the ability to deliver emotional education throughout the learning experience. It is only in this way that students can truly internalise and connect with the revolutionary spirit on a deeper level.

5. Reflections on the Deep Integration of Educational Tourisms

The composite nature of educational tourisms is most clearly reflected in the deep integration of the education industry, the broader educational sector, and the tourism industry, particularly with the growing integration trend across the entire industry chain. As a result, upstream and downstream participants in the educational tourism industry must adopt more innovative thinking, gain a deeper understanding of education, and engage more actively in sector-related activities. They can co-develop various products that provide intellectual and cultural value to diverse audiences through brainstorming and collaboration with partners along the educational value chain. The following six areas are of key importance to present more in-depth educational tourism content, foster more meaningful industrial integration, and offer better immersive learning environments.

The first is to promote the development of cross-sector educational tourism products. As the market continues to expand, schools, training institutions, travel agencies, educational tourism bases (or camps), and service providers are expected to gradually move toward integrated collaboration. The current fragmented market structure is likely to become more concentrated over time. For example, when a educational tourism base develops its products, it builds a new market recognition. By leveraging creative cross-sector integration, it can spark market interest, build a strong product reputation, and shape the perception of its target audience.

The second is to design educational tourism curricula based on scientific principles. In May 2017, the industry standard Educational tourism Service Specification (LB/T 054–2016), approved by the National Tourism Administration, officially came into effect. The document specifies that educational tourism products should be designed according to different school levels' characteristics and educational goals. This suggests that future curriculum development will emphasise thematic content and inquiry-based learning. Moreover, educational tourisms will no longer be limited to primary and secondary school students but extend to learners of all ages, supporting further market segmentation. Product design should begin at the source, aligning with users' cognitive needs. One practical approach is to involve educational and tourism departments, as well as residents, in the course design process. Based on the academic characteristics of each grade level in local schools, educational tourisms can be designed with different themes, durations, strong relevance, high quality, and distinctive features.

Third, the educational tourism supply system must be optimised. As educational tourisms are highly comprehensive activities, providers must not only upgrade their supporting facilities to meet the diverse needs of different learner groups, but also improve their online promotion platforms to stay in line with the ongoing digital transformation of society.

Fourth, the development of a qualified team of educational tourism instructors must be strengthened. Educational tourism instructors are a product of the integration between the education and tourism sectors. Educational tourism bases must ensure the quality of their instructors, as the instructors' professionalism directly affects the quality of course implementation. Therefore, efforts should be made to strengthen qualification assessments and instructors' monitoring and to carry out targeted training programs. By building a dedicated team of instructors, providers can ensure the delivery of high-quality educational tourism services.

Fifth, equal attention should be given to both learning and travel. Educational tourisms should avoid the imbalance of "all travel and no learning" or "all learning and no travel". Courses must be well-structured, thoughtfully integrating hands-on activities and reflective components to achieve a meaningful combination of travel and education. When organising educational tourisms, schools and travel agencies should carefully balance the academic seriousness of the "learning" aspect with the leisure-oriented nature of "travel", so that both can be enjoyed together. It is recommended that education and tourism authorities work together to refine the relevant regulations, using industry standards to guide key practices, operational procedures, and quality control, and to promote the healthy and standardised development of educational tourisms.

Sixth, scientific evaluation standards should be established for educational tourism bases. These bases should implement a robust evaluation system and build feedback platforms to understand users' real experiences better. By strengthening supervision and evaluation, they can ensure the service quality of the base itself and its educational tourism activities.

In the future, cross-sector integration will remain a mainstream trend in developing educational tourism products. Developing scientifically designed curricula will be key to shaping new market perceptions. Optimising the supply system is essential to meeting the diverse needs of educational tourism participants. Building a high-quality team of instructors is critical to ensuring the quality of education. Emphasising learning and travel is necessary for the sector's growth, and establishing scientific evaluation standards is vital for building strong brand recognition in the educational tourism industry.

6. Chapter Summary

This chapter has outlined the main models of "Educational tourism Plus" ("Educational tourism +") along with representative case studies. Analysing these cases has demonstrated how high-quality educational tourism products can help establish exemplary educational tourism bases and enrich the overall supply in the educational tourism market. As an emerging format that extends from the broader cultural and tourism sector, the educational tourism industry has gained significant popularity in recent years. It has been incorporated into primary and secondary school education plans. Educational tourisms serve as an essential entry point for implementing national policies related to experiential learning, representing a new model under the broader "Tourism Plus" framework.

The experiential nature of educational tourisms is steadily increasing, and a diverse "Educational tourism +" consumption structure is beginning to take shape. Although innovation within the "Educational tourism +" model will continue to evolve, it is essential to note that the base provides the foundation, but the content is the core. As a new type of tourism, the growth of educational tourisms depends on the physical infrastructure of educational tourism bases and on integrating innovative, meaningful content.

7. Post-chapter Exercise

1. What integration aspects should be considered in developing the "Educational tourism +" innovation model?
2. What key points should be considered in the "Educational tourism +" product development?
3. "Educational tourism +" is itself a form of interdisciplinary learning. Please provide an example and explain how it incorporates different subject areas.

Chapter 8

Market Research on Educational Tourism Marketing

Learning Objectives

1. Understand the types and content of market research.
2. Understand marketing research methods targeted at different segments of the educational tourism market.
3. Flexibly apply marketing research theories to the educational tourism industry.

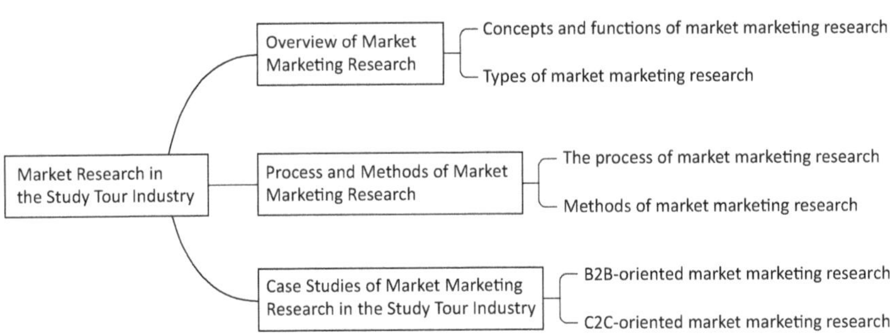

Learning Focus

1. Understand the general methods and processes of market research.
2. Understand the types and applications of marketing research in the educational tourism market.

3. Understand the methods used in the educational tourism market marketing research.
4. Understand practical marketing research approaches for different types of educational tourism markets.

Contents

1. Case Study

On 30 April 1999, in Camden, a small town in South Carolina with a population of around 8,000, a groundbreaking ceremony was held for a Haier production centre, which was established with an investment of 30 million USD by the Haier Group. Just over a year later, the first Haier refrigerator bearing the label "Made in the USA" was successfully produced. This marked the beginning of Haier's manufacturing operations in the United States, making it the first Chinese brand to manufacture and sell products directly in the U.S. market.

1.1. *Analytical questions*

1. Why did Haier choose the United States as the entry point for its international market expansion?
2. What types of research might Haier have conducted before entering the U.S. and European markets? What are some possible methods for carrying out such research?

2. Overview of Marketing Research

2.1. *The Concept and functions of marketing research*

Marketing research is systematically collecting, recording, organising, analysing, and reporting marketing information related to specific markets, products, or services. Its purpose is to identify various problems and opportunities, propose and evaluate potential marketing strategies, and assist decision-makers in making informed management decisions. Simply put, it is a structured and standardised process to support better decision-making. Marketing research emphasises the importance of providing systematic and reliable market information, making scientific and comprehensive research methods essential.

The educational tourism market is still in its early stages and involves various stakeholders such as government departments, educational institutions, students, and parents. It urgently requires thorough marketing research to clarify consumer needs and preferences.

Marketing research typically serves three main functions: descriptive, diagnostic, and predictive. The **descriptive function** involves collecting and analysing relevant information to present decision-makers with an overview of educational tourisms' internal and external market conditions, including consumer demand and competitors. It provides a factual description of the market, consumers, and the organisation. The **diagnostic function** focuses on interpreting, analysing, and comparing specific data and facts to evaluate a company's operational status and assist decision-makers in further action. The **predictive function** helps businesses make informed judgments about future market developments by analysing current conditions and identifying trends, thereby reducing the risk associated with decisions.

In recent years, due to real-world constraints such as the COVID-19 pandemic, many educational tourism enterprises have faced varying degrees of impact. As a result, internal diagnostics and market forecasting have become particularly important.

2.2. *Types of marketing research*

(1) **Classification by Research Content**
According to research content, marketing research can be classified into three primary categories: market demand research, research on controllable factors, and research on uncontrollable factors.

Market demand research centres on comprehending the maximum, minimum, and potential market demand, the characteristics and variations in demand for different products, and the market share held by competitors.

Research on controllable factors includes product, pricing, distribution channel, and promotional strategy research. Product research covers the characteristics of a company's products and competing products, market feedback, product structure, brand trademarks, packaging, and more. Pricing research investigates the elasticity of demand relative to price, how pricing changes for new and existing products affect sales, and how different types of costs influence pricing decisions. Distribution channel research evaluates current distribution channels' effectiveness, assesses intermediaries' marketing capabilities, and collects user feedback on these aspects. Promotional strategy research examines the effectiveness of personal selling, advertising, and other promotional methods.

Research on uncontrollable factors includes studies on the political environment, economic development, socio-cultural factors, technological trends, and competitors. For example, government policies on educational tourisms can significantly impact the industry's growth. In addition, differences in economic and cultural conditions across regions and among families indirectly influence consumer preferences for educational tourism products, particularly their acceptance of specific content and pricing.

(2) **Classification by Research Purpose and Method**
Based on research purpose and methodology, marketing research can be divided into four categories: **exploratory research**, **descriptive research**, **causal research**, and **predictive research**.

Exploratory research is conducted when the issues facing an enterprise or the market are unclear. Its purpose is to identify and clarify the nature of the problem and its underlying causes, as well as to understand external environmental factors related to the issue, thus laying the groundwork for further investigation. This type of research does not aim to produce definite answers, but rather to gather more information about the problem and potential solutions. Exploratory research often relies on secondary data sources, such as literature reviews and expert interviews. For instance, if a educational tourism company wishes to design products and travel plans encompassing all school levels but is uncertain about schools' expectations regarding educational tourisms, it would first conduct exploratory research. This might involve examining students' academic needs at different levels and reviewing relevant policies to establish a clearer direction for product development.

Descriptive research objectively describes and reflects existing market conditions while uncovering the internal relationships among various factors. It is typically conducted to depict the characteristics of a specific group of people or the features of a particular phenomenon. This research demands clearly defined objectives and aims to provide accurate, detailed descriptions. For instance, when launching various educational tourism products, we might investigate consumers' willingness to purchase different types of products and understand the demographic profiles of their target users. The research may reveal, for example, that high-end, international-oriented educational tourisms are especially appealing to teachers, students, and parents from foreign language schools. In contrast, public schools favour more versatile and cost-effective educational tourisms. Such analysis serves as an example of descriptive research. Unlike exploratory research, descriptive research relies more heavily on first-hand data collected through various research methods.

While descriptive research identifies correlations between two or more variables, causal research seeks to determine direct causal relationships among those variables. For example, if different types of schools demonstrate varying levels of acceptance towards premium and standard educational tourism products, and a educational tourism company wishes to investigate whether this is due to price differences, it might conduct causal research. This would involve testing whether increasing prices leads to declining acceptance among public schools, whilst keeping other variables constant. The primary method employed in causal research is experimentation.

Predictive research builds upon descriptive and causal research to forecast future market demand and company sales from both qualitative and quantitative perspectives. It serves as a crucial foundation for business planning and plays a vital role in guiding enterprise operations. For instance, if a company can identify the relationship between variables such as product price, demand, and school level through research, it may use regression equations to forecast sales volume or determine pricing strategies. However, real-world situations are often more complex, and the relationships among variables can be intricate. In the educational tourism industry, consumer demand is typically influenced by various factors, including government policy, product pricing, the duration of the educational tourism, and the destination. Therefore, it is often necessary to integrate multiple types of research—exploratory, descriptive, causal, and predictive—to ensure a comprehensive and accurate understanding.

3. The Process and Methods of Marketing Research

3.1. *The process of marketing research*

The process must be carefully planned and effectively implemented to ensure that marketing research achieves its intended results. Overall, the marketing research process can be divided into three main phases: the preparation phase, the implementation phase, and the conclusion phase. Each stage must be well connected to ensure clear objectives and efficient actions.

(1) **Preparation Phase**
The preparation phase involves identifying the research problem and objectives, and on that basis, developing a research plan. Researchers must first clarify the topic of investigation, either based on the needs of decision-makers or issues identified during market operations. Defining the research goal is the most critical step in research design. At this stage, researchers must specify the problem as clearly as possible, gather preliminary information, and conduct an initial analysis. Exploratory research often plays a key role during this stage. Once the issues are well understood, the research objectives can be

confirmed and the research plan can be developed. The research plan should clearly define the purpose of the research, the target population, the specific steps involved, the timeline, and the estimated costs.

(2) **Implementation Phase**

The implementation phase involves organising researchers to conduct in-depth investigations and collect data according to the research plan. Generally, if a company has a dedicated marketing research department, that department will lead the effort. If not, staff from other functional departments may collaborate on the research. Alternatively, the company may outsource the task to a professional consulting or research agency. Given the technical nature of marketing research, it is advised that professional personnel oversee the process to ensure quality and accuracy. Once the research team is confirmed, training must be organised to ensure all team members understand the research objectives, methods, and relevant legal or regulatory considerations.

During the data collection phase, the initial focus is on secondary data, which refers to existing information collected and organised for purposes other than the current study. This may include yearbooks, reports, official documents, academic journals, compiled collections, databases, and internal company records. Secondary data helps researchers understand the context of the issue, formulate hypotheses, and explore potential solutions. Such data can be obtained from internal archives, government websites, and public databases. However, all secondary data must be filtered to meet specific criteria: it should be accessible, relevant to the research topic, sourced from credible institutions, accurate, and up to date. Only secondary data that meets these standards can provide a valuable foundation for the research.

Despite being convenient and cost-effective, secondary data is often limited because it was not collected specifically for the current research purpose. Therefore, conducting fieldwork to obtain primary data is essential. Researchers must engage with target subjects during the scheduled period to collect original, first-hand information that addresses the research objectives. This is the primary source of data for the study.

(3) **Summary Phase of Marketing Research**

The data collected through marketing research must be organized, edited, analysed, and interpreted before it can become valuable information. **Data organisation and editing** involve coding, tabulation, and restructuring raw data into a more comprehensible format. **Analysis and interpretation** refer to statistical methods to further process the data, uncover underlying patterns, and ultimately present it in a form that supports conclusions.

Descriptive statistics are typically used in the initial data analysis stage, followed by more advanced methods depending on the research objectives. Throughout this process, all activities—whether organising, editing, analysing, or interpreting—must align with the research goals. It is essential to maintain objectivity and fairness when handling the data, clearly distinguishing between opinions and facts, as well as between causes and effects.

The final step in marketing research is often the preparation of a **research report**, which presents the research objectives, process, and outcomes in an integrated manner. The report serves both as a summary of the entire research process and as a critical reference for decision-makers in formulating strategies and plans. Therefore, all data collected during the research and the conclusions drawn from the analysis must be reflected in the report.

A typical research report comprises a title page, a cover, a table of contents, an executive summary, an introduction, research methodology, research findings and results, study limitations, conclusions and recommendations, appendices, and references. To ensure the report is accessible to decision-makers, the content should emphasise the research objectives and findings, forging connections between theory and practice. Technical discussions of data analysis methods ought to be simplified as much as possible to facilitate understanding.

3.2. *Methods of marketing research*

The use of research methods to obtain data is a critical component of the marketing research process. Secondary data is primarily collected through literature review, while primary data is mainly obtained through inquiry-based research, observational research, and experimental research.

(1) **Inquiry-Based Research**

Inquiry-based research refers to a method in which the researcher poses a series of structured questions to respondents and gathers data based on their answers. This method is widely used in marketing research because it allows researchers to obtain information about respondents' attitudes and intentions toward a product or service. Depending on the mode of interaction between the researcher and the respondent, inquiry-based research can be divided into three types: personal interviews, telephone interviews, and questionnaire surveys. Personal interviews involve one-on-one, face-to-face communication with the respondent. They can be further categorized into in-home interviews, where an appointment is made to meet at a specific location, and intercept interviews, where respondents are approached randomly in high-traffic areas such as plazas, stores, or school entrances. Telephone interviews are conducted via phone and are valued for their timeliness. Questionnaire surveys were traditionally conducted by mail, but with the development of the internet, online questionnaires have become increasingly popular. To ensure the effectiveness of the results, the design of questionnaires must be scientifically sound.

Inquiry-based methods are also widely used in marketing research for educational tourism programs. For example, researchers might randomly select participants for brief interviews during an event or distribute satisfaction surveys to parents and students after the activity to understand their satisfaction levels and collect suggestions for product and program improvement.

Regardless of the specific method used, it is essential to minimize sampling errors and ensure that the sample is representative so that the results are accurate and reliable. In addition, thorough preparation and planning of interview questions are required. Personal interviews, in particular, demand a higher level of skill from the researcher—those who can respond flexibly to various reactions from interviewees are more likely to reduce bias and improve data quality.

(2) **Observational Research**

Observational research refers to the method of collecting research data by directly observing and recording the behaviour of research subjects or specific phenomena. This method places great emphasis on the objective recording of what is observed. Compared to inquiry-based methods, observational research is generally more objective, but the data obtained can be more limited in scope. When using observational research, it is essential that the subjects or settings being observed are observable, recurring, and occur frequently enough for timing and frequency to be recorded accurately. Observational research is particularly suitable for collecting precise data, such as pedestrian traffic in a specific area during a certain period, or customer behaviours leading up to a purchase decision.

There are various types of observational research, with two of the most common being mystery shopping and trace observation. Mystery shopping involves assigning researchers to pose as customers and enter either the company or a competitor's establishment to gather data. This method requires a high level of professionalism from the researchers. Trace observation, on the other hand, gathers data by examining physical traces left by customers—for example, comments recorded in a customer feedback book in a retail store.

Observational methods are used less frequently in the context of educational tourism marketing research due to the disconnect between the actual buyers (parents) and the participants (students). Most observation-based research in this sector focuses on student behaviour, although some parent-child educational tourism products provide a broader range of observable subjects. As the educational tourism industry continues to grow, a more mature system for observational research is expected to gradually emerge.

(3) **Experimental Research**

Experimental research refers to how professionals use experimental techniques to study the relationships between variables. It focuses on explaining causal relationships under specific conditions to provide more scientific support for decision-making. The premise of experimental research is to define the research variables, including independent, dependent, and controlled variables. To minimise error, the research conditions must be tightly controlled. Generally, experimental research falls into three categories: laboratory experiments, field experiments, and computer simulation experiments.

Although experimental research effectively identifies causal relationships, it is often challenging to implement. Experimental methods are not widely or systematically employed in the context of educational tourism marketing research. They are more commonly used to forecast product pricing and sales performance; however, numerous variables often complicate the establishment of standardised causal links.

4. Case Study of Marketing Research in the Educational Tourism Industry

4.1. *Marketing research targeted at distribution channels*

In general, camps, bases, or travel agencies engaged in the educational tourism business obtain their customers mainly from direct consumers (C-end), schools, or other intermediary channels, such as travel agencies, educational training institutions, media platforms, and community-based organisations. While many camps and bases today can directly collaborate with schools to secure student participation or use their own media platforms to recruit students, intermediaries still play an important role in acquiring customers in bulk. Their significance at the current stage is reflected in two key aspects.

First, local governments in many provinces and municipalities still encourage schools to partner with travel agencies to organise educational tourism programs. In practice, schools rely on travel agencies to coordinate services such as meals, accommodations, attractions, transportation, and educational tourism instructors. Furthermore, due to prior working relationships and habitual decision-making, many schools choose trusted travel agencies or qualified educational tourism organisations to handle their programs, even if alternative options exist.

Second, the educational tourism market is still highly fragmented by region, making cross-regional sales challenging. For example, if a secondary school in Shaanxi Province wants to organise a educational tourism in Zhejiang, it would rarely contact a Zhejiang-based educational tourism agency or camp directly. Instead, it would more likely commission a local travel agency in Shaanxi. In such cases, for a Zhejiang-based educational tourism provider to acquire the school as a client, it must build a cooperative relationship with a travel agency in Shaanxi and treat it as a key distribution channel. Additionally, many non-study-tour-specific organisations, such as media outlets, educational institutions, and youth centres, hold valuable, yet untapped, student resources that could be converted into participants. Therefore, conducting marketing research targeted at distribution channels remains necessary and valuable for many camps, bases, and travel agencies.

(1) **Key Considerations in Channel-Oriented Marketing Research**
Channel-oriented marketing research typically involves several categories of questions.

The first category concerns the **evaluation of intermediary quality**: From which intermediaries can we acquire student sources? Which intermediaries are suitable for us? Which are effective? Which are the ones we are more likely to build partnerships with? What are the development costs for each intermediary? What are the maintenance costs and commission rates? What services must we provide to facilitate cooperation with each intermediary?

The second category involves **research on the intermediary's student base**: Where does the intermediary's student base come from? What is the size of their current or potential student base? What is the purchasing power of these students? Is the relationship between the intermediary and its student base stable? What is the student conversion rate? For example, some media outlets are able to consistently recruit tens of thousands of participants annually for summer and weekend camps, while some travel agencies may have volatile student sources, relying on attractive products and competitive pricing to secure cooperation with schools.

The third category involves **existing partnerships**: Which educational tourism institutions, camps, or bases is the intermediary currently working with? What is the status of those partnerships? Are the relationships stable? Are the channel fees high?

The fourth category concerns **the intermediary's product offerings**: Does the intermediary or the end customer control product selection? What products is the intermediary currently promoting, and which have they previously promoted? What are the pricing structures and sales performance of these products? Which products are performing well, and what advantages do they have? How well do our products align with the intermediary's portfolio? Is the intermediary willing to promote our products?

The fifth category looks at the **intermediary's future development strategy**: What is the intermediary's current or future attitude toward cooperation in the educational tourism sector? Are there any planned strategic adjustments that could impact their involvement in the educational tourism market or their partnership with us?

The sixth category involves **feedback from existing intermediaries**: Is the intermediary promoting our products appropriately? Are they satisfied with our offerings? What suggestions or demands do they have for improvement? Are their students willing to purchase our products, and why or why not? What feedback or requirements have their students provided regarding potential product improvements?

(2) **Common Methods Used in Channel-Oriented Marketing Research**
Due to the particular characteristics of intermediaries, large-scale surveys or experimental research are generally unsuitable for marketing research in this context. Instead, **inquiry-based methods** or **observational methods** are commonly used. These include on-site interviews, telephone interviews, questionnaires, and on-the-ground observation of intermediaries' marketing activities to gather relevant information.

In addition, for intermediaries with strong existing partnerships, joint research can be conducted to understand better the characteristics and needs of the student base associated with each channel.

For example, a educational tourism organisation based in Eastern China conducted a market research project following the 2019 spring educational tourism season. Since December 2018, the organisation has introduced six independently developed educational tourism products to travel agencies in the southern market, specifically in Guangdong, with stable partnerships with local schools. During the spring season of 2019, over 3,400 students from nine schools, including Guangzhou No. 1 Middle School, Shunde Le Liu Middle School, and Dongguan Songshan Lake School, participated in the company's educational tourism programs.

The purpose of the research was to evaluate the sales and promotion performance of these travel agency channels during the spring season, identify shortcomings in the products and services, guide improvements for the fall educational tourism offerings, and estimate potential sales for the upcoming season. After multiple rounds of internal discussion, the following research plan was developed:

Research Period: June to July 2019.

Research Subjects: Key persons responsible for the educational tourism divisions or primary business contacts from 25 medium to large travel agencies in four cities: Guangzhou, Dongguan, Foshan, and Shantou.

Target Market Area: Major cities in Guangdong Province.

Research Team: 1 staff member from the Product Department and four from the Marketing Department.

Research Objectives: To comprehend how various travel agency channels have promoted the organisation's products; to identify the factors that influence agencies' willingness to promote and schools' readiness to adopt the products; to assist in future product development and pricing decisions; and to evaluate the current status of agency partnerships while determining their willingness for continued cooperation to support sales planning.

Research Methods:

Secondary data was gathered from literature and online resources, encompassing details on each travel agency's overall business volume, historical and current seasonal cooperation, and partnerships regarding educational tourisms.

Primary data was collected using inquiry-based research methods (e.g. interviews) and limited observational research to gather more direct insights.

Interview Outline for the Marketing Research

1. Overall Promotion of Spring Educational Tourism Products
   a. In what form, and by whom, are educational tourism products typically promoted to schools by your agency?
   b. In the spring of 2019, how many schools did your agency visit to promote educational tourism products, and among them, how many received promotions for our products specifically?
   c. What is your opinion of our educational tourism products? Why are you or your colleagues willing—or unwilling—to recommend them?
   d. What kind of educational tourism products would make you very willing to recommend them to schools?
   e. What changes should we make to our educational tourism products to increase your willingness to recommend them to schools and other clients?

2. Adoption and Acceptance of Educational Tourism Products
   a. While promoting our educational tourism products, how many schools showed interest? How many ultimately chose our products?

b. Were there schools that originally intended to adopt our educational tourism products but later changed their minds? What were the reasons?
c. In your opinion, do schools that are uninterested in our products feel that destinations like Zhejiang, Jiangsu, or Shanghai are unappealing, or do they find the products themselves lacking in appeal?
d. Could you summarize the main issues schools have raised regarding our educational tourism products? What improvements do you think we should make to increase adoption?

3. Overall Channel Strategy and Educational tourism Investment
   a. What are your company's current core business areas?
   b. What proportion of your company's business is accounted for by educational tourisms? Is it a primary business area? How does your company view this business segment?
   c. How many staff members are currently working in the educational tourism division? What are their roles? Are they dedicated full-time to educational tourism activities? How much of their time is allocated to this work?
   d. Do you think the return on investment in your company's educational tourism division is high?
   e. For the upcoming fall educational tourism season, will your company increase, reduce, or maintain the current level of investment in this area?
   f. What adjustments can we make to help your company improve efficiency while reducing workload?

Research Results
Among the 25 selected travel agencies, 17 successfully completed in-person interviews, 4 completed online communication, and four did not participate in the study. This resulted in 21 valid samples. For details, see Table 8.1.

Findings Summary

According to the survey results, of the 21 travel agencies interviewed, 10 agencies included the organisation's educational tourism products in their official product brochures. They promoted these products to schools

Table 8.1. Survey questions and high-frequency responses.

Survey Questions	High-Frequency Responses
Reasons why the educational tourism product was not included in the promotion plan	1. The educational tourism product was combined with activities such as visits to entrepreneurship companies and internet firms, which are not aligned with the primary purpose of educational tourisms for schools in East China. 2. Although the product had relatively good packaging, it was still priced at 200–400 RMB higher than other local products in East China, resulting in limited market appeal. 3. None of the travel agencies had previously promoted this educational tourism product.
Reasons why schools selected the educational tourism product	1. The educational tourism product was new and attractive. 2. Schools agreed with the educational philosophy behind the product. 3. The quality of the instructors provided for the educational tourism was considered high.
Reasons why schools did not select the educational tourism product	1. There were doubts about the logistical execution and the guidance ability of the educational tourism instructors. 2. The product was considered too "specialized" and not suitable for schools seeking more comprehensive educational tourism options. 3. The product was relatively expensive.

through their retail outlets or educational tourism departments. Five agencies advertised the products individually by creating posters or product leaflets for schools. The remaining six agencies either promoted the products verbally or did not formally introduce them to schools.

Ultimately, four travel agencies successfully secured deals with nine schools. Of these nine schools, four made their selection based on long-term cooperative relationships with the travel agencies, grounded in mutual trust. In comparison, the other five schools selected the product through a bidding process led by parent committees.

Regarding overall strategy and investment from intermediary partners, four travel agencies indicated they would not develop their educational tourism business in the short term due to internal business planning. In comparison, the remaining 17 agencies expressed strong intentions to expand their involvement in the educational tourism market. The survey also revealed that Guangdong Province has raised the credit requirements for high school comprehensive practice courses, thereby bolstering travel agencies' confidence in promoting educational tourism products in the near future.

Post-survey Review

This research project was largely completed as planned. Following the survey, the organisation promptly adjusted its products, upgraded its services, and launched coordinated marketing campaigns. As a result, during the autumn 2019 educational tourism season, more travel agencies began recommending the organisation's products to schools, and a growing number of schools showed interest. Ultimately, the organisation's revenue from the autumn educational tourism season in the southern market increased by more than 30% compared to the spring season.

(1) **Case Study: School-Oriented Marketing Research**

Schools are key stakeholders in educational tourisms. Marketing research targeting schools focuses on understanding their specific needs for educational tourism products and collecting feedback. The research subjects typically include school administrators, teachers, students, and their parents, which means various research methods must be used. Interviews are usually employed when engaging with school administrators, while questionnaires are more commonly used for teachers, students, and parents.

Take, for example, a research project conducted by an institution for a themed educational tourism titled "Pillars of the Nation, Tech-Innovation in Shanghai", designed for high school students. The research plan was as follows:

Research Period: September 2019.

Target Respondents: Educational tourism coordinators from 20 high schools.

Target Region: [Name of region omitted].

Research Objective: To understand high school needs for educational tourism programs and gather suggestions on specific product itineraries.

Research Method and Process:

1. Market research using inquiry-based methods. The research team visited each of the 20 schools and conducted in-person interviews with the educational tourism coordinators to gather insights into each school's educational tourism planning, product preferences, desired destinations, and learning objectives.

From the interviews, it was found that school coordinators expected educational tourism products to achieve several goals, such as broadening students' horizons, exposing them to modern scientific and technological advancements, improving understanding of industrial development, enhancing safety awareness and emergency response capabilities, and providing motivational exposure to university-level education.

Based on these findings, Shanghai emerged as the preferred destination for most respondents. Accordingly, a thematic educational tourism route titled "Pillars of the Nation, Tech-Innovation in Shanghai" was designed, leveraging high-quality local resources such as the Shanghai Science and Technology Museum, Yangshan Deepwater Port, Shanghai Jiao Tong University, and the Shanghai Public Safety Education and Training Base.

4.2. *Marketing research targeted at the C-End*

The C-end market, or the individual customer market, is considered one of the most challenging segments in the broader educational tourism and camp education industry. **Product positioning** is the core element when developing educational tourism products for the C-end market. Precise product positioning and rich content are essential to attracting C-end customers, and **word-of-mouth marketing among customers** plays a vital role in customer acquisition in this segment.

Therefore, C-end marketing research typically consists of two key parts:

(1) **Pre-experience questionnaires** are conducted before customers participate in the product, and
(2) **Satisfaction surveys** conducted after the product experience. Both are primarily carried out using **questionnaire-based inquiry methods**.

When designing the questionnaires, it is crucial to consider the distinction between **product users (students)** and **paying customers (parents)** in the educational tourism market. The questionnaire content must allow for independent feedback from both parties to gather more valid and comprehensive information.

As for implementation, C-end market surveys are typically conducted **on weekends or public holidays**, with locations such as libraries, museums, and children's activity centres—places where parent-child families are concentrated—being ideal. If a educational tourism organization operates its own camp or experience centre in a city centre, such venues are especially suitable for conducting C-end marketing research.

(1) **The Core of C-End Marketing Research: Product Positioning**

Educational tourism products targeting the C-end market are often designed as summer camps or weekend camps, where the campers themselves play a central role in the operation of each camp session. Marketing is an activity that should take place year-round, serving as the most direct means of recruiting both participants and staff.

During the marketing process, organisations must consistently communicate the general value of the camp experience and the unique content of their offerings to define each camp session's positioning clearly. Marketing expert Dr. Dwight Jewson has emphasised that "the most advanced marketing strategies begin with defining who we are". Camps must ensure that their product positioning resonates with their target customers, or at least appeals to them, and then align all aspects of the program around that clearly defined identity.

(2) **Content of C-End Marketing Research: Customer Benefits**

Customer benefits are a key focus in marketing research for the educational tourism industry. In the C-end market, educational tourism product providers must analyse the benefits they can offer potential customers. These core values or benefits should be listed clearly and systematically for a particular camp session or program module.

Potential customers often compare the stated benefits of one program with those of others. Therefore, when describing these benefits, the messaging must be clear, concise, and compelling enough to capture customer interest. The features of summer camp services and their corresponding customer benefits are illustrated in Figure 8.1.

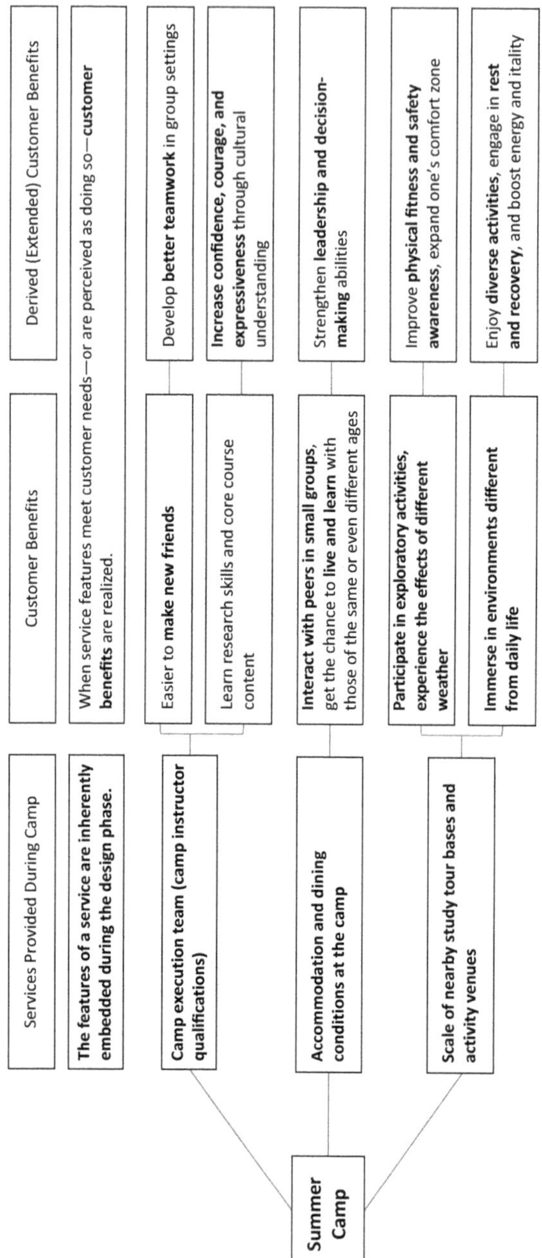

Figure 8.1. Features of summer camp services and corresponding customer benefits.

In 1990, Maguire Associates, an educational consulting firm, conducted a study revealing the five most valued characteristics of camps among parents. These were: friendly and caring counselors, enjoyable camp activities, overall program quality, safety, and personalised attention given to each camper.

(1) **Questionnaire Research—A Key Method in C-End Market Marketing Research**

Questionnaire research is often the preferred method when analysing how customers perceive the benefits they receive. Typically, surveys are distributed to current campers and their parents, with the content focusing on areas where their expectations were met and where they were not. The timing of when the questionnaire is given to campers also affects the quality of the research. For example, distributing the survey immediately upon the camper's arrival or sending a digital version long after the camp ends may result in low response rates and less reliable feedback. Therefore, questionnaire research targeting the C-end market should be conducted as soon as possible after the camp session concludes, while the experience remains fresh in the campers' minds. This timing allows participants to provide more insightful feedback. Additionally, it is not advisable to pressure campers to complete the questionnaire, as that could impact the quality of their responses.

While conducting market research, marketers should also promptly update their current marketing strategies and materials, such as the product detail pages on WeChat official accounts and mini-programs. It is essential to carefully review every aspect of the marketing touchpoints—from website information, brochures, and promotional videos to camp materials and the scheduling of marketing events. This includes checking whether the content aligns with the target audience, whether it is accurate and engaging, and whether the benefits of the camp experience are clearly communicated and easily recognised.

5. Chapter Summary

Marketing research is inherently a standardised method and activity for acquiring the information necessary to support marketing decisions. It is guided by a well-developed methodology and bolstered by practical cases in the market. As an emerging industry, the educational tourism sector still lacks extensive experience in marketing research. Furthermore, owing to the industry's unique characteristics, marketing research strategies must be customised to various consumer groups, each with distinct priorities. This chapter draws on classic marketing research cases in the educational tourism industry and integrates relevant theoretical frameworks to explore professional methods and procedures for conducting marketing research. The aim is to contribute to the standardisation and formalisation of marketing research practices within the educational tourism sector.

6. Post-chapter Exercise

1. What are the commonly used research methods in educational tourism marketing?
2. Analyse the differences between B-end and C-end marketing research.
3. Try designing a educational tourism product demand survey questionnaire targeted at students from public schools.

Chapter 9

Formulation and Implementation of Research Travel Marketing Plans

Learning Objectives

1. Understand the basic concept of a marketing plan and become familiar with its contents and formulation steps.
2. Master the key points in formulating a research travel marketing plan.
3. Master the key points in implementing a research travel marketing plan.

Knowledge Framework

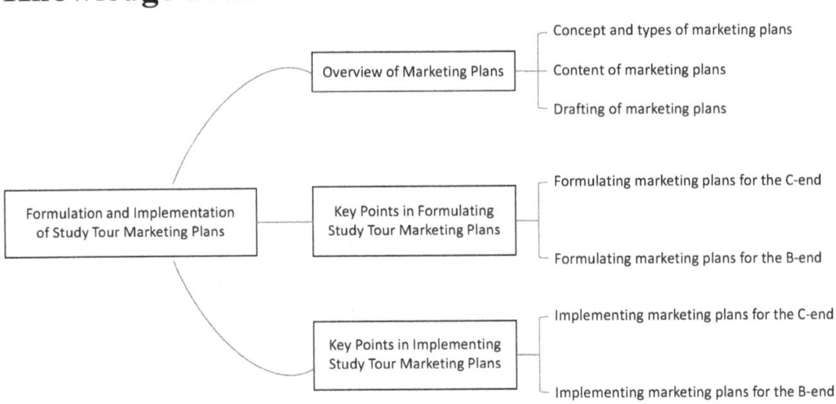

Learning Focus

1. Understand the content and formulation steps of a marketing plan.
2. Understand the application of marketing plans in the educational tourism industry.
3. Master the key points in formulating and implementing marketing plans for different segments within the educational tourism market.

Contents

1. Case Study

In the autumn of 2021, Hailiang Educational tourisms invited principals from various schools under Hailiang Education Group to participate in a unique educational tourism activity titled "Principals' Great Love". In this event, Hailiang Educational tourisms organized for the principals to experience the educational tourism products firsthand as if they were students, gaining multi-dimensional and comprehensive insights into the courses, venues, and execution processes involved. The activity received unanimous praise from the principals, and the related educational tourism products were subsequently included in the educational tourism plans of their respective schools.

1.1. *Analytical questions*

1. In essence, can Hailiang Educational tourisms' "Principals' Great Love" activity be considered a marketing activity?
2. What aspects need to be considered when planning and implementing a educational tourism marketing activity?

2. Overview of Marketing Plans

2.1. *The concept and types of marketing plans*

The marketing plan is one of the most crucial components of the marketing process. After conducting marketing research to analyse the internal and external environment of an enterprise, it is necessary to formulate a marketing plan to establish clear marketing objectives and determine the strategies, measures, and steps required to achieve them. For a company, a marketing plan represents "doing things right". It helps managers clearly define the company's development direction and prioritise relevant actions. The implementation and progress tracking of a marketing plan ensure that the company's marketing efforts are carried out in an orderly and structured manner.

Marketing plans can be classified into different types based on various criteria. In terms of time span, marketing plans can be divided into long-term, medium-term, and short-term plans. A long-term plan is a strategic document, usually spanning more than five years. A medium-term plan typically covers a period of more than one year but no more than five years. A short-term plan usually refers to a one-year cycle, such as an annual plan. Based on the scope involved, marketing plans can be divided into overall and specific marketing plans. An overall marketing plan refers to the comprehensive plan for all marketing activities of the enterprise, while a specific marketing plan focuses on a particular product or issue, such as a new product marketing plan, brand plan, or channel plan.

2.2. *The content of a marketing plan*

The content of a marketing plan mainly consists of three parts. The first part is a situational analysis, which describes the current operating environment of the enterprise. This information is primarily obtained through marketing research and answers the question "Where are we now?"

The second part emphasises the enterprise's future development goals, answering the question "Where are we going?" The third part outlines the specific actions needed to achieve these goals, including how to allocate human, material, and financial resources, as well as how to implement and monitor the plan. Overall, a marketing plan must clearly define its expected goals, most directly in terms of the anticipated economic benefits for the enterprise. This ensures that all operational actions remain focused on achieving the intended objectives. In addition, a marketing plan should identify the resources required for implementation, helping the enterprise to plan costs in advance, allocate personnel accordingly, and ensure that responsibilities are clearly defined. This promotes better internal coordination and supports the enterprise's continued development.

Specifically, a complete marketing plan typically consists of the following components: situational analysis, marketing objectives, marketing strategies and action plans, marketing budget, and marketing control.

(1) **Situation Analysis**

Situation analysis primarily examines the marketing environment in which the enterprise currently operates, focusing on key factors that affect business development, such as the market, products, competitors, and macro-environmental elements. It begins with identifying the needs of a specific target market and uncovering the characteristics of both existing and potential consumers. While analysing the status of one's own products, attention should also be paid to the strengths and weaknesses of competitors to help the enterprise anticipate their actions and better develop its own competitive advantage. At the same time, it is essential to consider the macro-environment—namely, the overall marketing environment—as understanding policy directions and socio-economic and cultural development trends can help the enterprise better identify growth pathways. Attention to new technologies also assists businesses in breaking through traditional marketing barriers. On this basis, a more detailed assessment of the opportunities and risks faced by the enterprise should be carried out.

(2) **Marketing Objectives**

Marketing objectives are the core of an enterprise's marketing plan, and their formulation relies on situational analysis. These objectives include financial goals—profit-oriented targets that must be achieved upon task completion, composed of indicators such as profit margin, sales volume, and return on investment. Financial objectives must be translated into marketing objectives to be effectively realised. Marketing objectives are measured by indicators such as market share, number of distribution outlets, and price levels. The objectives must be clear and specific rather than vague, measurable during implementation, and also account for time constraints. While focusing on corporate profitability, the development of marketing objectives should also take into account the interests of consumers and society, aiding the enterprise in managing this relationship effectively. Therefore, in addition to direct marketing profits, enterprises should include in their objectives the enhancement of brand image, the exploration and accumulation of marketing experience, and the construction of an optimal marketing system over the coming period.

(3) **Marketing Strategy and Action Plan**
To achieve marketing objectives, corresponding marketing strategies and action plans must be formulated. A marketing strategy refers to a broad course of action, requiring a clear identification and selection of the enterprise's target market and decisions on market positioning and strategies concerning product distribution channels, pricing, and promotion. An action plan refers to the detailed scheme for implementing various marketing strategies—clarifying what is to be done, when, by whom, and with what approximate cost. It is recommended that the entire action plan be presented in tabular form, specifying the start and end dates of each step, task requirements, responsible personnel, costs, and marketing expenses. Marketing strategy and action plans complement each other and jointly ensure the execution of marketing objectives.

(4) **Marketing Budget**
The marketing budget refers to the optimal amount of funds required for executing marketing strategies and allocating budgets across various marketing plans. It is a refinement of the marketing objectives and strategies and should directly reflect its role in supporting these strategies. Additionally, it serves as a crucial basis for coordinating the work of different departments. The marketing budget may constitute a substantive projected profit and loss statement, estimating the total marketing revenue in the income column and projecting the total costs and marketing expenses in the expenditure column, ultimately deriving the expected profit. Furthermore, the marketing

budget acts as the standard by which the company evaluates the performance of its marketing department.

(5) **Marketing Control**
Marketing control refers to the examination and regulation of the marketing plan. Its primary function is to monitor the progress of the marketing plan and ensure that its implementation remains on schedule and its outcomes are evident. To guarantee the effective execution of marketing control, both marketing objectives and budgets should be broken down into stages, ensuring that progress at each phase can be tracked and substantiated. Generally, if it is found that a certain phase of the marketing plan has not been completed, the causes must be promptly analysed, and corrective measures must be proposed to ensure that the overall marketing objectives can still be achieved. Therefore, it is essential to define the procedures for marketing control and implement an effective supervisory mechanism.

2.3. *Formulation of the marketing plan*

(1) **Principles for Developing a Marketing Plan**
A well-constructed marketing plan is a critical safeguard in achieving an enterprise's marketing objectives. When formulating a marketing plan, an enterprise should adhere to the following three principles: First, the plan should always be based on the enterprise's overall marketing objectives and be consumer-centred, ensuring alignment with the strategic direction. Second, full consideration must be given to the enterprise's internal and external environments. Thorough marketing research and situation analysis are essential to clarify the opportunities and risks the enterprise faces. A lack of vital information from internal and external environment studies may result in short-sighted marketing strategies and deviations in marketing objectives. Third, the marketing strategies and action plans within the plan must be clear, specific, actionable, and capable of being effectively monitored and evaluated through marketing control.

(2) **Steps in Formulating a Marketing Plan**
Generally speaking, the formulation of a marketing plan involves the following steps: First, a situation analysis is conducted through marketing research to determine the marketing objectives, which

subsequently guide the following procedures. At this stage, decision-makers often need to confirm these objectives to ensure that the strategies and goals are feasible. Second, a draft of the marketing plan is developed. This process involves collaboration across departments and requires that the tasks outlined in the plan be specific and clearly defined. Once the draft is complete, it is submitted to the relevant departments for discussion. Should there be any objections, revisions must be made within a specified timeframe. Finally, once the official marketing plan has been finalised, the enterprise must implement it in an orderly manner and carry out all associated marketing control measures to ensure the achievement of the marketing objectives.

3. Key Points in Formulating a Educational Tourism Marketing Plan

When formulating a marketing plan for educational tourism programmes, it is essential to consider the unique characteristics of the industry fully. The participants of educational tourism products are primary and secondary school students, while the decision-makers and payers are the parents. Schools are also involved in the selection of educational tourism products and suppliers, as well as in the supervision of the implementation process. In addition, the role of intermediary channel partners often needs to be taken into account during the actual marketing process.

Therefore, the formulation of a marketing plan for educational tourisms should fully consider the needs and concerns of all parties involved in the project. Generally speaking, students—as the end users—tend to value the enjoyment and playability of the educational tourism product; parents are more concerned with cost-effectiveness and safety; schools primarily focus on safety and educational value; and intermediary channel partners are chiefly interested in profit margins and positive reviews.

3.1. *Formulating a marketing plan for the C-end market*

When formulating a marketing plan aimed at the C-end market, it is essential to analyse the existing consumer groups, investigate demographic information within the area served by the product, understand similar competing programmes, and evaluate the current marketing and service status. These elements constitute the foundational data for creating a marketing plan. Relevant personnel should invite both the staff who will implement the plan and management team members—possibly even relevant consumers—to participate as consultants.

When developing the marketing plan, the first step is to re-examine the marketing message and clearly define the intended audience. Reviewing the overall marketing goals and positioning is crucial, as clarifying the product's positioning constitutes a vital aspect of the planning process. Some products are positioned as comprehensive experiences designed to expose participants to a variety of programmes, while others may be tailored to a specific niche market. If the product is intended to serve different audiences, information about its positioning may need to be communicated through various means, employing different marketing tools for each group.

At present, most educational tourism enterprises in the market recognise that traditional product brochures alone are inadequate. There is a need to develop additional marketing tools, such as websites and video platforms, as primary means of reaching target audiences. Regardless of the tools used, it is essential to ensure that the marketing message directly targets the intended group and consistently communicates the product's mission, expected outcomes, and positioning.

To reduce costs, some educational tourism organisations frequently consolidate all information into a single brochure. However, whether this approach can effectively convey relevant marketing information to various groups using the same facilities, such as summer camp participants, parents, rental clients, weekend programme customers, day campers, and residential campers, remains questionable. Using the same information to market to diverse groups may lead to confusion. Therefore, the best method is to create dedicated brochures tailored to each target group, ensuring the marketing messages are more precise.

Furthermore, websites promoting educational tourism products should feature clear and prominent sections for various target groups (e.g. students, parents, staff, etc.) on their homepage. It is also essential to assess whether the current marketing tools reflect those of other companies and to make comparisons to identify the tools that should be implemented moving forward. Marketing tools generally categorise into four types: individual interaction marketing, material-based marketing, advertising marketing, and public relations marketing. Balancing these four types of tools can help mitigate the limitations of any single approach.

3.2. *Formulating a marketing plan for the B-end market*

When formulating a marketing plan for the B-end market, it is essential to focus on demand and segment the target market and consumer groups. Differentiated marketing plans should be adopted for different regions to accommodate varying policy requirements and market conditions. For instance, in areas where educational tourism policies are clearly defined and governmental support for educational tourism development is strong, marketing plans may focus on the school end to secure bulk semester-based educational tourism orders. In contrast, in regions with vague policies and insufficient governmental support, the marketing plan may prioritise the channel end, leveraging the resource advantages of intermediaries to promote summer camps, weekend camps, and other types of educational tourism products. Within the same region, marketing plans should also be tailored to different consumer groups. For example, in the public school sector, marketing efforts should focus on the cost-effectiveness and educational value of educational tourism products. In contrast, for private schools, marketing should highlight the uniqueness and enjoyment of the programmes.

For the B-end market—especially bulk educational tourism activities organised during the school term—product structure should be a key focus in the marketing plan. A systematic and professional product structure can enhance schools' recognition of educational tourism offerings and facilitate long-term, ongoing partnerships, providing staged, growth-oriented educational tourism courses for schools and students across different year levels.

Various marketing channels are also vital in developing a marketing plan for the B-end market. These channels include training institutions with age-appropriate student sources, scenic spots, venues, and campsites with their own traffic, along with companies that provide engineering, equipment, teaching aids, or logistical services in long-term collaboration with schools. The focus of the marketing plan will differ depending on the type of marketing channel. For training institutions, the plan should prioritise how closely the educational tourism product aligns with the channel's existing content. In most cases, tailored product solutions must be designed for the user base of the channel to convert training students into educational tourism participants, while also enhancing customer retention and revenue for the training institution. For scenic spots, venues, and campsites, the marketing plan should emphasise how to persuade the channel to provide traffic support and how to execute traffic conversion. Negotiations often revolve around the exchange of traffic resources or custom-developed courses. For companies with access to school resources, the marketing plan should highlight the unique advantages of the educational tourism products and the professionalism of the delivery team to alleviate the concerns of the channel. It should also clearly articulate the profit margins and cooperation models for the educational tourism projects.

Additionally, pricing strategies must be planned for B-end marketing initiatives. This involves two key aspects: Firstly, an overall marketing pricing system should be established based on region and product line, ensuring that the same product adopts a uniform pricing structure within a given region. Secondly, the pricing framework for channel partners should also be standardised in advance, ensuring that the same product series is offered to similar types of intermediaries at broadly identical price levels. Market guide prices should be provided to avoid excessively high or low final transaction prices. If certain partners intend to sell at significantly higher prices, the conventional channel products must be custom-designed or improved to create distinction, thereby maintaining the stability of the market pricing system across the entire channel.

4. Key Points in the Implementation of a Educational Tourism Marketing Plan

Execution of a Marketing Plan refers to the process of transforming the marketing plan into concrete tasks and ensuring the goals are achieved as scheduled. It involves addressing the questions of "who", "when", "where", and "how". To implement a marketing plan effectively, the first consideration must be its feasibility—facilitating the transition from planning to execution.

In the implementation of a educational tourism marketing plan, the first step is to convert the plan into actionable tasks. It is essential to clarify what tasks need to be accomplished at each stage of the plan, who is responsible for them, when and where they will be carried out, and what objectives are to be achieved. With this clarity, the tasks can then be effectively completed. Ideally, a project team should be established, with a project manager responsible for refining and breaking down the overall goals, assigning the right personnel, detailing specific tasks, and managing the entire project schedule.

Secondly, it is necessary to provide essential support for the execution of these tasks, which includes personnel support, resource support, and process control. Personnel support means allocating a sufficient number of appropriate staff based on the task requirements and ensuring that they have adequate time to complete their assignments. Resource support means that for any significant marketing project, the company's decision-makers must act as key guarantors, so that resources and assistance can be promptly obtained when challenges arise. Process control involves holding weekly or appropriately scheduled project meetings to track progress across different components, preventing deviations in objectives, timelines, or direction. If any issues are identified, they must be promptly addressed and rectified.

Finally, it is important to conduct a timely review and summary, analysing the marketing campaign's strengths and shortcomings. The findings should be archived and organised to provide valuable insights and references for future marketing activities.

4.1. *Implementation of marketing plans targeting the C-end*

Given the unique nature of C-end products, the implementation of marketing plans in the C-end market is often closely linked to the institution's scheduling of educational tourism programmes. In fact, in most cases, the marketing plan for the C-end market is essentially aligned with the institution's programme schedule. Typically, the schedule of a programme is directly determined based on market demand.

(1) **Channels for Publishing C-End Market Marketing Activities**
The classification of marketing activities observed in both domestic and international markets is broadly similar and includes, but is not limited to, alumni return days, camper reunions, weekend camp activities, family days during the programme, camper family gatherings, telemarketing activities, programme delegation recruitment, and invitation-based experience camps. The channels for publishing marketing activities and product information typically take the following forms: distribution of brochures and flyers, referrals from returning campers, dissemination of short online videos, and creation of news stories.

a. **Alumni Return Day**

With the exception of certain special cases, the primary source of enrolment for C-end programmes is returning campers. Most programmes significantly depend on their past participants, and the repurchase rate among parents serves as a key indicator of programme quality. According to the 2018 China Camp Education Industry Report, the average repurchase rate for camp education products is 49%. Clearly, the lower the re-enrolment rate of returning campers, the more challenging the recruitment task becomes for camp directors; attracting new campers typically requires considerably more time, effort, and money.

Therefore, the first priority for camp directors is to focus on returning campers. It is vital to inform them in advance of the summer camp schedule, specific dates, and special events, for example, through official WeChat posts. Once a returning camper registers, related content, such as website information and promotional posts, can be tailored to reflect this. Many camps have found that regularly sending newsletters or stories from the programme to campers and their parents is a great way to maintain connections, deepen emotional bonds, and encourage re-enrolment. In particular, such content should include photos or mentions of key instructors. While producing such materials does not require significant cost, they must be inspiring and engaging—mentioning past campers by name, sharing photos from previous camps, and recalling memorable experiences can be especially effective.

Some camp companies have also created WeChat chat rooms, bulletin boards, and blogs so that campers can stay connected beyond the camp period. Establishing online communities is an effective and low-cost method for keeping everyone interested in the camp engaged, fostering relationships, and maintaining access to information.

b. **Camper Reunions**

Camper reunions often rekindle enthusiasm for the camp among returning campers and camp counsellors. These gatherings, held annually, offer an opportunity to revisit stories from past programmes, browse photos and videos, and relive the joy once experienced. A typical camper reunion includes two key elements: interaction with former camp instructors and the presentation of past camp photos or video footage. Such reunions help campers remember the positive experiences they had and strengthen the connections among them.

Camp organisations should be adequately prepared with promotional materials for the upcoming programme and bring them to the reunion to encourage early registration for the next session.

If most campers come from various provinces (or even countries), institutions may organise regional reunions. These events tend to be larger in scale, generate greater excitement among campers, and offer more social opportunities for both campers and staff. Regional camper reunions also allow for more personal engagement between camp leaders and participants, making it easier to invite prospective campers and parents who have shown interest in the programme. During the reunion, returning campers may be encouraged to recommend new participants (or staff members), and the organisation can follow up after the event. Some camps even encourage campers and staff to bring friends along to the reunion.

Some educational tourism companies host slideshows on their websites or video platforms featuring a particular programme, showcasing relevant photos, videos, and camper feedback. These can serve as a form of "visual camper reunion", enabling those who live far from other campers to reminisce about their experiences. Former campers can also use the website to share pictures and stories with friends, eliminating the inconvenience of coordinating visits to the camp in person.

c. **Weekend Camp Activities**

If the camp is not far from the city centre, the camp director may organise public activities during off-season periods such as autumn, winter, and spring, outside the main summer camp season. Whether it is a one-day excursion, a camping event, or a weekend hiking trip, these occasions provide ideal opportunities to introduce summer camp programmes. If a group is renting the site during such an event, the camp can also promote its upcoming summer sessions open to individual participants. Furthermore, the camp director is responsible for confirming and managing relevant photos, videos, and other promotional materials from these activities.

Currently, many general educational tourism and camp education companies operate under a light-asset model. This implies that numerous campsites are only affiliated with a specific camp brand during the actual programme period. However, this type of business partnership does not prevent institutions from running weekend programmes aimed at C-end customers. In fact, even when a brand-oriented marketing approach is the primary business model, institutions can still provide a series of structured courses based on their curriculum during the w

The key lies in identifying and focusing on the core customer benefits—offering high-quality, immersive, and content-rich experiences that can be delivered in a day or over a short period. These products serve as a practical extension of the overall marketing plan. By doing so, the organisation can improve brand visibility and provide a more systematic presentation to the C-end market, thereby maximising the impact and value of weekend camp marketing activities.

(2) **Key Considerations for Designing C-end Marketing Activity Registration Forms**

C-end marketing activities often involve the distribution of registration forms to potential consumers. These forms are crucial for collecting consumer information and must therefore be designed with care. Depending on how the registration forms are archived and how frequently they are utilised, one may choose from various formats such as a card, a full-page form, a half-page form, or an online registration form. When determining the layout of the form, it's also important to consider whether the information needs to be entered into a computer database. Most importantly, the data collected through the registration form will serve as initial information for individual or group participants in the database, so thoughtful consideration must be given to what information may be needed in the future.

Registration forms for individual participants may include the following details:

1. Address (postal codes can indicate regional information);
2. Home telephone and mobile number;
3. Email address;
4. Date of birth;

5. Gender;
6. School and current or upcoming academic year (state the current year at registration or the year to be entered in the upcoming autumn term);
7. Name of parents or guardians and emergency contact details;
8. Workplace addresses, email addresses, and telephone numbers of each parent or guardian;
9. Height and weight;
10. Religious affiliation (especially when religious activities are part of the programme);
11. Preferred group mates in the same living group;
12. How the participant learned about the programme;
13. Other organisations' programmes or activities the participant has attended;
14. Any special needs requiring attention, whether physical, psychological, or mental;
15. Previous outdoor camping or camp experience;
16. Cancellation and refund policies.

Usually, the above information is followed by a parental or guardian authorisation statement. The parent or guardian must sign this section to confirm their consent for the participant to attend the camp and agree to the payment terms. Only the participant's signature is required for adult participants to confirm their participation and fee payment.

The authorisation statement may include:

1. Permission for the organisation to use photos or videos of the participant taken during the camp;
2. Declaration of household income if the participant qualifies for local or national funding;
3. Details of the participant's family or personal insurance provider (particularly when the camp does not provide health or accident insurance, or when family insurance will serve as the primary or secondary coverage);
4. Medical or surgical authorisation statements (required by some hospitals for emergency care);
5. Information regarding the collection of minors (listing individuals authorised to pick up the child besides the parents);
6. Dietary requirements.

(3) **Scheduling Tool for C-end Marketing Plans: Gantt Chart**
The Gantt Chart, also known as a horizontal bar chart, utilises bars to visually represent the progress of tasks, projects, and other time-related developments, reflecting how these elements evolve. It is named after its creator, Henry Laurence Gantt. The Gantt Chart is an exceptionally suitable tool for institutions to use in scheduling both product timelines and marketing plans. It allows staff across various departments, such as instructors, marketing, branding, and customer service, to coordinate their efforts in advancing the camp sessions and marketing tasks. This ensures all personnel are aligned and progressing according to the same timeline within the company's overall work.

4.2. *Implementation of B-end marketing plans*

When implementing marketing plans targeted at the B-end (business clients), the foremost consideration is the particular nature of the target audience. These are typically institutional leaders, heads of organisations, or even school principals. As such, special care must be taken when selecting execution personnel, venues, and timing.

Firstly, the staff responsible for implementation must be experienced, or at the very least, possess sufficient interpersonal charm and the ability to adapt quickly. Secondly, outreach methods may include individual visits or formal invitations. If inviting B-end stakeholders to an event, the venue must be chosen due to their professional status. Thirdly, scheduling should be carefully planned. For example, when visiting principals and academic directors of private schools, one should avoid the end-of-term period and instead arrange meetings at the beginning of the term, about a week after exams, or during the term at a mutually convenient time.

The following is an example of how Hailiang Research Travel implemented its marketing (promotion) plan for the "School Curriculum System Development" project, illustrating key points in B-end marketing execution.

1. **Team Formation**
As the "School Curriculum System Development" marketing project involved research travel activities throughout an academic term, Hailiang attached great importance to it and adopted a project management approach. A senior leader with project management experience was appointed as the project manager. The project team

was divided into several groups: Product, Curriculum, Branding & Communication, Marketing, and Operations. Each group had an appointed team leader who reported directly to the project manager.

- The **Product Team** was responsible for improving the Hailiang research travel curriculum system, preparing and updating products accordingly.
- The **Curriculum Team**, working closely with the Product Team, was tasked with developing any new lessons or teaching processes required due to curriculum adjustments.
- The **Branding & Communication Team** externalised the Product and Curriculum designs into marketing materials.
- The **Marketing Team** was responsible for contacting and inviting school principals, assembling visiting groups, conducting on-site school visits, establishing long-term relationships, and completing promotional and sales tasks.
- The **Operations Team** supported the Product and Curriculum teams at the execution level by providing information and ensuring readiness for project implementation.

2. **Plan Breakdown**
   Upon receiving the marketing plan, the project manager convened the group leaders and finalised the project plan after three rounds of meetings. A detailed Gantt chart was created, breaking the plan into 52 specific tasks. Each task had a designated person-in-charge, a supervisor, a final deadline, and acceptance criteria.

3. **Execution Assurance and Process Management**
   First, the project manager held weekly meetings to review progress and assess whether the project was on track regarding objectives, direction, and deliverables. If issues were identified, timely adjustments were made to ensure the project stayed on course and met its timeline.

   Second, each task had an assigned person in charge and a supervisor, ensuring accountability and quality control.

   Finally, as the stakeholder representing business needs, the company's general manager regularly received progress updates from the project manager and directly participated in project meetings. This allowed for the prompt identification of challenges and the provision of necessary resources, ensuring the final project outcomes aligned with business expectations.

5. Chapter Summary

After conducting marketing research to understand the educational tourism market's overall demand and strategic expectations, the next step is to develop a marketing plan based on established goals to ensure their practical realisation. In this process, the formulation and implementation of marketing strategies and action plans will vary in focus depending on the specific market segments targeted.

This chapter, grounded in fundamental marketing theory, examines the key considerations for creating and implementing marketing plans aimed at various target groups within the educational tourism market. The objective is to support the holistic development of the educational tourism industry by ensuring that marketing efforts align well with audience needs and strategic goals.

6. Post-chapter Exercise

(1) What steps are involved in formulating a educational tourism marketing plan?

(2) What are the everyday marketing activities for the B-end and C-end educational tourism markets, respectively?

(3) Try to design a marketing plan for a summer camp product targeting the C-end market.

Chapter 10

Exploring the Application of Smart Marketing Platforms in Educational Tourisms

Learning Objectives

1. Understand the structure of the smart marketing platform system for educational tourisms.
2. Learn the design methods of smart tools in educational tourism marketing.
3. Master the design methods of the smart evaluation system for educational tourisms.

Knowledge Framework

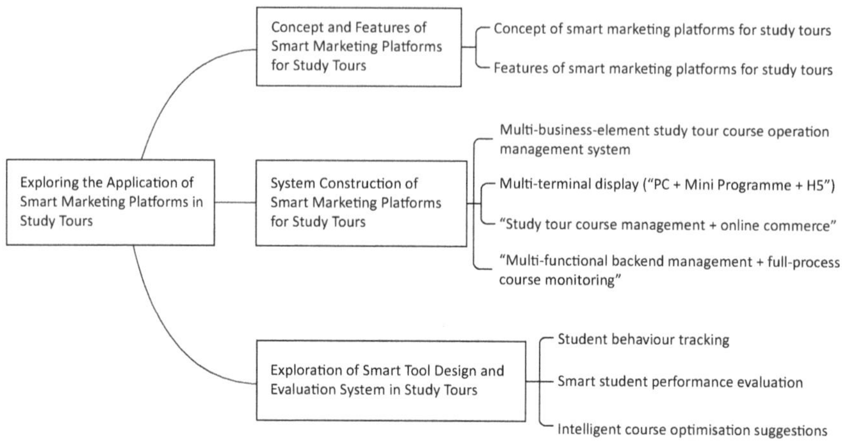

Learning Focus

1. Master the design methods of smart tools and smart evaluation systems in educational tourism marketing.
2. Understand the structure of the smart marketing platform system for educational tourisms.

Contents

1. Case Study

On 18 June 2021, founder and CEO of Yueyang Education, Dr Zhou Lin, was invited by the Department of Culture and Tourism of Henan Province to deliver a special lecture titled "Constructing a Smart Educational tourism Camp System" for a educational tourism training programme.

Dr Zhou first interpreted the far-reaching implications of the new Gaokao (college entrance examination) reform and its significant impact and practical demands on the educational tourism industry. He then analysed the policy background, social significance, and industry demand for building an innovative educational tourism camp system. Following that, he introduced the Yueyang Intelligent Camp Management System, which comprises seven layers, four frameworks, and eleven system functions:

1. **Seven Layers:** User layer, terminal layer, spatial layer, application layer, basic support layer, data resource layer, and infrastructure layer.
2. **Four Frameworks:** Standards and regulations framework, safety assurance framework, operations and maintenance framework, and data analysis framework.
3. **Eleven System Functions:** Course scheduling system, educational tourism evaluation system, dormitory management system, asset management system, attendance system, base management system, archive management system, activity resource management system, information release system, central fund management system, and decision analysis system.

Dr Zhou also presented the core architecture of the Yueyang Smart Educational tourism SaaS Platform. The platform offers educational tourism institutions a free OMO (Online-Merge-Offline) tool to enhance the professionalism of their courses (by providing pre- and post-tour lessons aligned with in-school curriculum objectives under the OMO model), enable smart service and management, and help institutions break out of the challenges of low-frequency and low-margin operations.

Through its "three terminals" (teacher terminal, parent terminal, and PC management terminal), the SaaS platform assists educational tourism organisations in improving course quality and service standards. It provides high-quality pre- and post-tour lessons and personalised educational tourism reports, allowing parents to stay informed of their child's journey anytime. It improves operational and workflow efficiency, enabling complete online management of itineraries, routes, and personnel, and allowing for one-click room and transport assignments. It also strengthens user retention and conversion through multiple parent touchpoints, various retention strategies, and diversified conversion channels to maximise revenue.

Educational tourism institutions can use this platform across "ten aspects": "before, during, after, display, evaluate, ensure, assign, monitor, manage, profit".

1.1. *Reflection and discussion questions*

1. Based on the case, what functions should a smart educational tourism marketing platform include? Can they be aligned with the actual stages of running a educational tourism? Share your thoughts.
2. Does your city have a similar smart educational tourism marketing platform? If yes, please share your findings in class.

2. Concept and Features of the Smart Marketing Platform for Educational Tourisms

2.1. *Concept of the smart marketing platform for educational tourisms*

The smart marketing platform for educational tourisms is a comprehensive intelligent system solution developed in response to the needs of the educational tourism industry. It integrates smart learning, online and offline marketing, online services, and big data management. Utilising internet technologies, it provides technical support for educational tourism operations and management, enabling the digital display of infrastructure, service facilities, and course content at educational tourism bases (or camps). It also streamlines the operational processes of these bases and visualises educational tourism courses through computer technology. Furthermore, it enhances the evaluation methods of educational tourisms using big data technology, allowing real-time online recording of students' participation and overall performance during educational tourisms.

2.2. *Features of the smart marketing platform for educational tourisms*

(1) **Digitalised Business Entities**
A major feature of the smart marketing platform is its ability to maintain comprehensive data records. All entities involved in the business process are digitally documented in a real-time and accurate database. For example, it creates digital records of all individuals (or organisations), items, and events related to a educational tourism, forming a relational database. This ensures that the entire process is traceable, with all business data centralised on one platform for easy access and input.

(2) **Automated Business Processes**
Once business entities are digitised and operated online, the platform automates processes that were traditionally manual and repetitive, such as order processing and descriptions of educational tourism scenarios. Tasks like course listing and delisting, customer order-driven services, and activity management can be fully automated through designated application functions. Offline operations are also enhanced through the integration of software and hardware—for example, in ticketing, entry validation, and broadcast control—minimising labour costs while boosting operational efficiency and profitability for educational tourism enterprises.

(3) **Data-Driven Decision Making**
Data-driven decision-making is a standardised operational approach of the platform. It relies on big data and cloud computing to support managerial decisions for educational tourism enterprises. This encompasses decisions related to the direction and scale of course development investment, marketing expenditures, pricing and commissions, financing activities, diagnosing operational quality

issues, recruitment, and personnel management. The information collected through big data enables more accurate forecasting of market trends and assists decision-makers in making more informed choices, thereby increasing the likelihood of success.

3. Overview of the Smart Marketing Platform System for Educational Tourisms

The smart marketing platform for educational tourisms is an integrated intelligent system solution developed specifically to meet the needs of the educational tourism industry. It combines smart learning, online and offline marketing, online services, and big data management. Leveraging emerging technologies such as cloud computing and the Internet of Things (IoT), it operates via the internet and connected network terminals to support the entire educational tourism process—before, during, and after the experience.

The system aims to broaden students' horizons, enrich their knowledge, deepen their connection with nature and culture, and increase their exposure to collective living and public ethics. Ultimately, it enhances the added value of each component in the educational tourism journey—food, accommodation, transport, activities, learning, and research. The platform enables proactive sensing, information interconnectivity, intelligent decision-making, and automated content delivery.

The smart marketing platform developed for the Yichang camp is a representative example of such system construction and is illustrated in Figure 10.1.

The smart marketing platform for educational tourisms plays a role on three key levels:

1. **Industry Level**—It enhances the strength and influence of the educational tourism industry while elevating the educational value of educational tourisms.
2. **Demand Level**—It improves the overall experience for primary and secondary school students, allowing them to feel the convenience brought by technology and increasing the fun and engagement of educational tourisms.

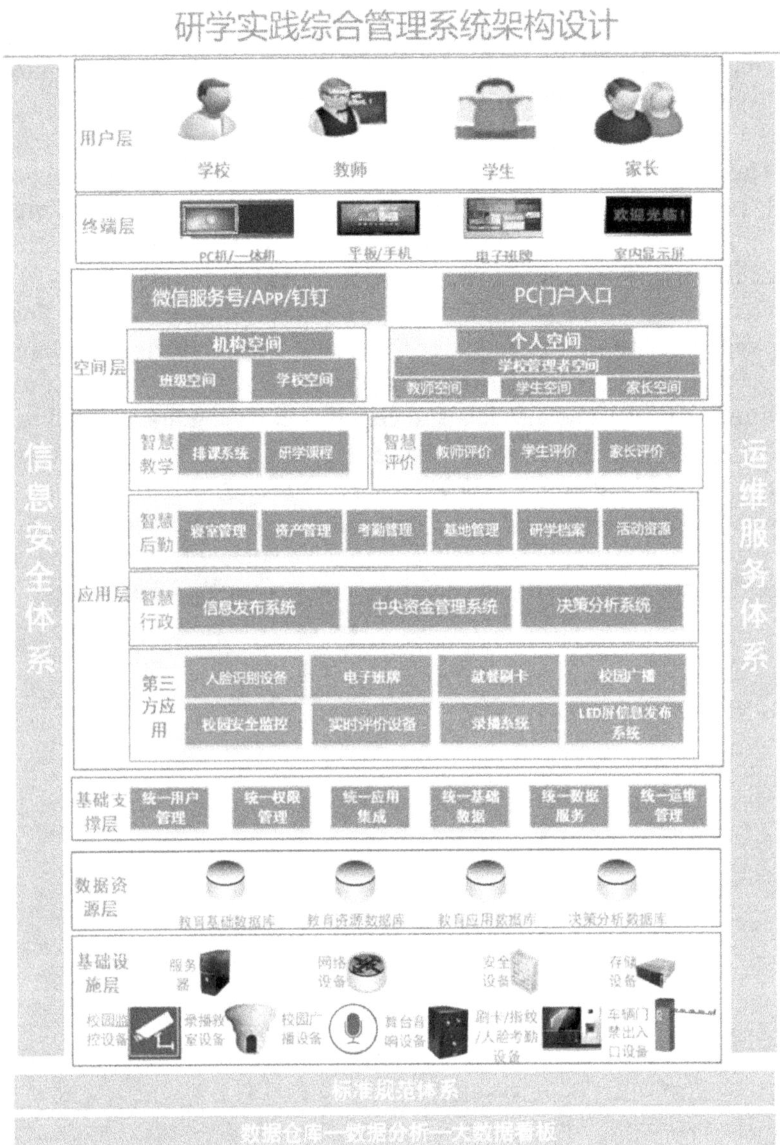

Figure 10.1. Yichang camp smart marketing platform system architecture.

3. **Service Level**—It strengthens stable cooperation between educational tourism enterprises and intermediary service providers, and enhances the promotional value of scenic spots and educational tourism bases.

The platform operates based on four core elements: **Smart Learning, E-Marketing, Intelligent Services, and Business Management**. Through the collaborative operation of these four elements, the platform builds its overall framework and practical operation model.

Taking the **Chongqing Primary and Secondary School Social Practice Education Management Service Platform** as an example (see Figure 10.2), the platform covers the education administrative departments of various districts and counties in Chongqing as well as all social practice education bases. It connects with all primary and secondary schools in Chongqing, enabling effective management of courses, activities, evaluations, and resources related to social practice education. The platform also provides timely and reliable social practice course services, establishes individual social practice records for students, and

Figure 10.2. Chongqing primary and secondary School social practice education management service platform.

offers accurate data to support the evaluation of students' overall competencies.

Chongqing's primary and secondary school social practice education efforts include the development of social practice curricula, the construction of education bases, and the establishment of a fully digital and networked management platform to oversee the entire process of student activities and educational tourisms. The **Chongqing Primary and Secondary School Social Practice Education Management Service Platform** covers over 400 social practice education bases across the city. Its users include students, teachers, base administrators, district- and county-level managers, and municipal-level managers. The platform is deployed at the municipal level and applied across five levels of users. It manages all behaviours and data associated with students' participation in social practice activities—including video, images, audio, and text records—ensuring that the data is real-time, authentic, and reliable.

Backed by big data, the **Smart Marketing Platform for Educational tourisms** supports flexible product combinations and booking options, offers predictive guidance on travel and related costs, and introduces social interaction functions for users. It enables real-time data mining for interactive engagement, navigation during travel, hotspot guidance, traffic prediction at educational tourism bases, and personalised recommendations for study-tour-related commercial services. The standard configuration of the educational tourism smart marketing platform mainly includes the following four elements.

3.1. *Multi-merchant cluster educational tourism curriculum management system*

The Smart Marketing Platform for Educational tourisms uses blockchain and AI technologies to build a big data platform tailored to the educational tourism industry. It creates an intelligent ecosystem connected to educational tourism bases, educational tourism organisations, tour instructors, user credit modelling, smart destination marketing, order forecasting, user behaviour analysis and prediction, secondary consumption guidance, and safety dispute handling.

The platform adopts a model of "merchant onboarding plus platform self-operation", which meets the online operational management needs of multiple merchants, as shown in Figures 10.3 and 10.4.

Figure 10.3. Multi-camp support (Each Camp Managed Separately via Its Own Backend).

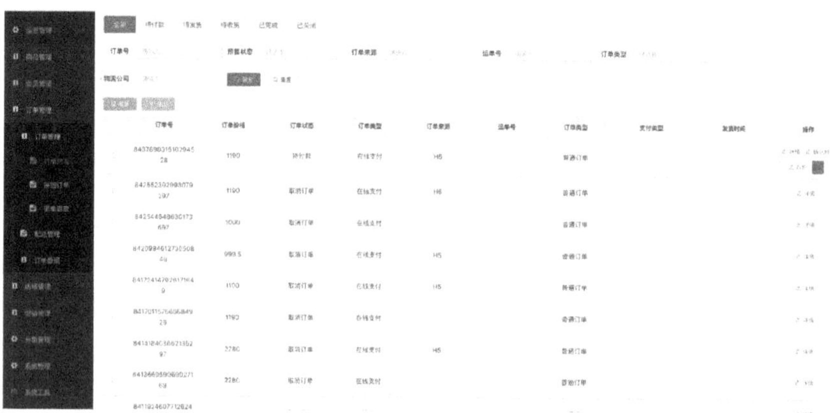

Figure 10.4. Camp Session registration and order management.

1. Multi-Port Display ("PC + Mini Programme + H5")
   The Smart Marketing Platform for Educational tourisms adopts a framework-based development approach with code separation and clean design. This allows for one-time development and multi-terminal use, supporting deployment across mini programmes, H5 pages, official WeChat accounts, and other access points, as shown in Figure 10.5.

(a)

(b)

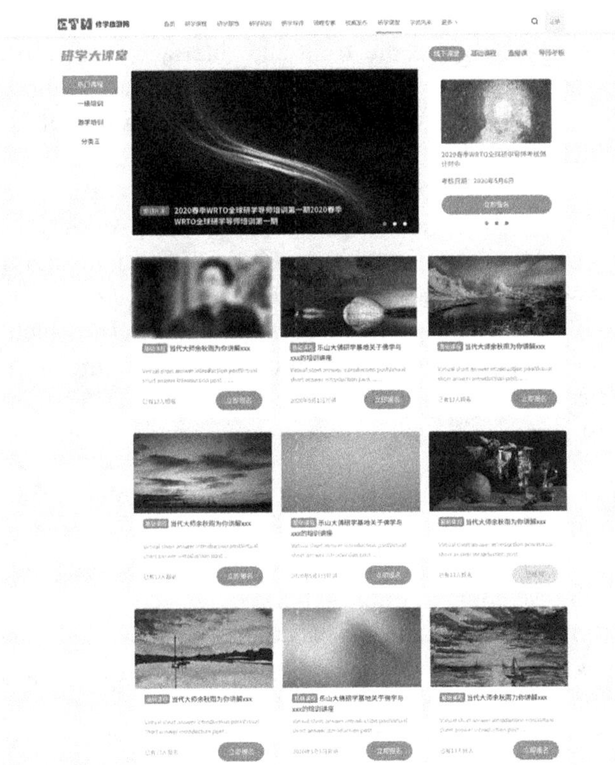

(c)

Figure 10.5. Multi-platform display.

2. "Educational tourism Course Management + Online Store"
   Educational tourism courses are uploaded to the smart marketing platform by tour bases and agencies. Designated school administrators select courses via the platform. Educational tourism bases may review and approve the selected courses. District and county education committees can view and approve schools' course selections in real time. The municipal education committee can monitor course selection and implementation data across regions and educational tourism bases. Meanwhile, parents can make unified purchases through the educational tourism online store, thereby achieving intelligent course management, as shown in Figure 10.6.

3. "Multi-Function Back-End Management + Full-Course Chain Management"
   Ordinarily, we only see the front-end interface of a website, while access to the back-end is reserved for administrators with authorised accounts. So, what management functions does a website's back-end system include?

   The multi-function back-end management system comprises three main areas: administrator management (also known as access control), which allows for adding new administrators and modifying their passwords; database backup, a feature designed to ensure the security of user data; and file upload management, which manages images and other files uploaded by users when adding products.

   Full-course chain management covers several aspects, as illustrated in Figure 10.7.

   (1) *Financial Management*: This mainly involves the addition, modification, and management of products, as well as the approval of those products. The page for the asset management section is shown in Figure 10.8.
   (2) *Settings Management*: This section allows for modifying various types of enterprise information. The page for the role management section is shown in Figure 10.9.
   (3) *Resource Management*: This section enables viewing detailed order information and the status of order processing.
   (4) *Data Management*: This section allows viewing, editing, and deleting member profiles. It also includes functions to lock or

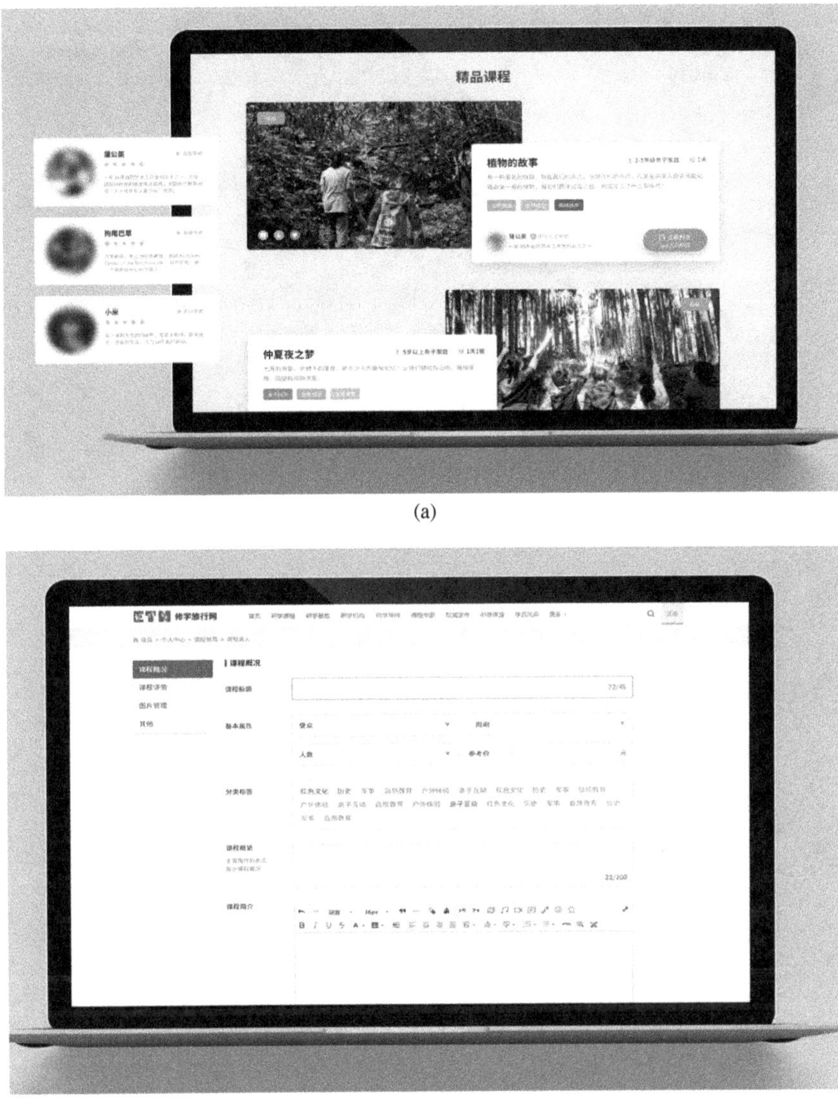

(a)

(b)

Figure 10.6. Educational tourism course online management and procurement.

unlock accounts and enables real-time communication with members. The page for the data analysis section is shown in Figure 10.10.

Figure 10.7. Full-link course management.

Figure 10.8.

(5) *Course Management*: This section supports the uploading of courses and completing course information for educational tourism bases and organisations.

(6) *Marketing Management*: This section manages feedback and user messages. Registered members' messages can be replied to

Figure 10.9. Role management.

Figure 10.10. Data analysis.

online, while messages from non-registered users can be answered via the online messaging function. The page for the marketing centre section is shown in Figure 10.11.

Figure 10.11. Marketing centre.

4. Exploration of Smart Educational Tool Design and Intelligent Evaluation System for Educational Tourisms

Smart educational tools for educational tourisms are not only designed for teachers but can also serve as materials for children to engage in their own "growth work". These tools are capable of assessing the effectiveness of a child's educational tourism experience. Therefore, when designing the level of operational difficulty, it is essential to follow a logical progression—from simple to complex, from easy to difficult, and from concrete to abstract. This approach ensures that children of varying abilities can use and operate the tools without feeling discouraged due to limitations in their capabilities. At the same time, smart educational tools can be integrated with aspects such as students' physical and mental health, on-site alerts, and post-tour evaluations, as illustrated in Figure 10.12. This integration enables educational tourism organisations and coordinators to make more informed, data-driven decisions, while also providing participants with a richer and more meaningful educational tourism experience.

1. Student Behaviour Monitoring
   The research learning monitoring system based on student behaviour data includes monitoring equipment, data collection devices, and data

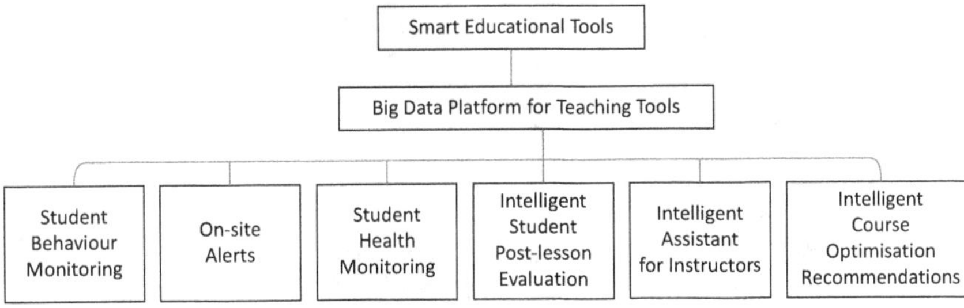

Figure 10.12.

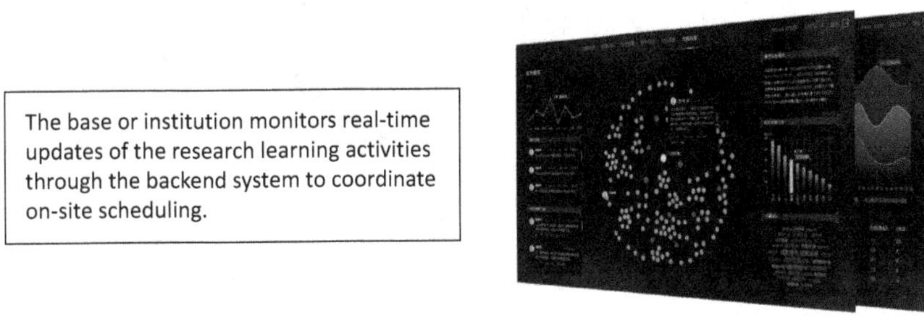

Figure 10.13. Student behaviour monitoring.

transmission devices. The monitoring equipment consists of a primary communicator, a primary processor, a memory unit, and an output device. Both the collection device and the primary communicator are connected to the data transmission device, while the primary communicator, memory unit, and output device are linked to the primary processor. The collection device transmits the monitored data to the primary communicator via the data transmission device. The primary processor then sends the classroom performance information, derived from the student behaviour data, to the output device. The student behaviour data is obtained from the data monitored by the research learning monitoring equipment and the data model pre-stored in the memory unit, as illustrated in Figure 10.13.

2. Intelligent Student Post-Trip Evaluation
   The establishment of an intelligent student post-trip evaluation platform serves two primary purposes: first, to enhance the innovative competitiveness of educational tourism products, and second, to focus on guidance and encouragement. Students and their parents jointly

Figure 10.14. Intelligent student post-trip evaluation.

participate in educational tourism experiences. Parents can provide genuine feedback on their child's performance through this platform, while students can share their honest impressions of the educational tourism base (camp). In this process, students are both the main participants in the experience and the subjects of self-evaluation, as illustrated in Figure 10.14. This approach helps stimulate student engagement and promotes harmony between teaching and learning through evaluation. In short, establishing a smart, user-friendly, visualised, and objective evaluation platform ensures that users of educational tourism courses truly benefit.

3. Intelligent Course Optimisation Recommendations
   Educational evaluation involves making value judgments about academic activities, processes, and outcomes based on specific

educational values or goals. It utilises feasible scientific methods to systematically collect, analyse, and organise information, aiming to improve education quality and provide a basis for educational decision-making.

A tripartite evaluation system can be introduced for designing an innovative evaluation system or intelligent learning tools for educational tourisms. This system involves experts from the tourism and education sectors collaboratively developing evaluation standards. It integrates data from students, parents, and tutors to create a comprehensive big data set for educational tourism evaluations.

This process not only enhances the development of educational tourism courses and improves educational tourism bases but also optimises the tutor teams. Moreover, it addresses professionalism, convenience, and timeliness issues when tutors evaluate students, as illustrated in Figures 10.15–10.17.

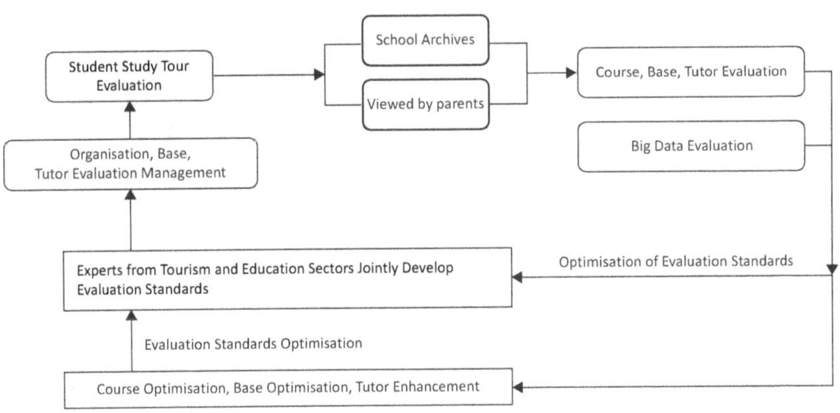

Figure 10.15. Development and optimisation of educational tourism evaluation standards.

(a)

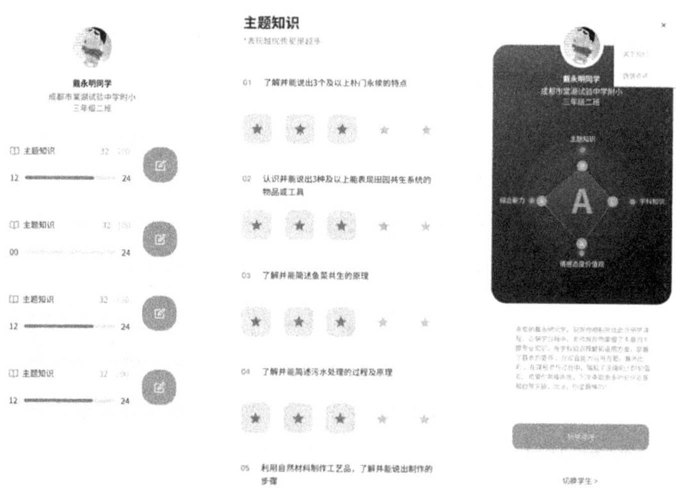

(b)

Figure 10.16. Tripartite evaluation system.

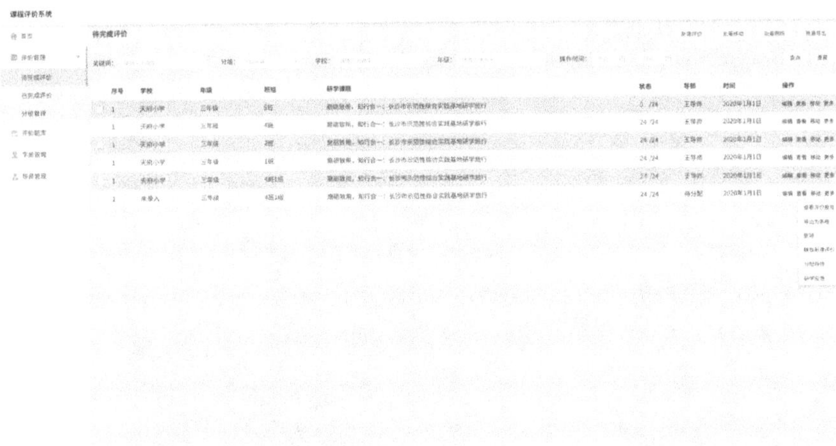

Figure 10.17. Management courses and related evaluations.

5. Chapter Summary

The construction of smart marketing platform systems for educational tourisms occupies a vital position in the educational tourism market, as online platforms overcome the limitations of time and space and offer advantages in big data, traffic, convenient management, and distribution channels. These platforms assist regulatory authorities in managing and monitoring the educational tourism industry, helping educational tourism institutions enhance their course quality and service standards. Moreover, smart marketing platforms allow parents to stay updated with exciting developments in their children's educational tourisms systematically, enabling evaluations to be effectively integrated into the educational tourism experience. This creates a systemic advantage that supports the sustainable development of the educational tourism industry.

6. Post-chapter Practise

1. What management systems generally make up a smart marketing platform for educational tourisms?
2. In your opinion, what factors should be taken into account when designing smart teaching tools for educational tourisms?

Chapter 11

New Media Marketing in Educational Tourisms

Learning Objectives

1. Understand the concept and characteristics of new media in educational tourisms.
2. Understand the operational strategies and work responsibilities of new media in educational tourisms.
3. Understand that creative content is the core of new media marketing for educational tourisms.
4. Understand online public opinion management in the context of new media for educational tourisms.
5. Learn how to build a personal educational tourism brand and use new media to market that brand.

Learning Framework

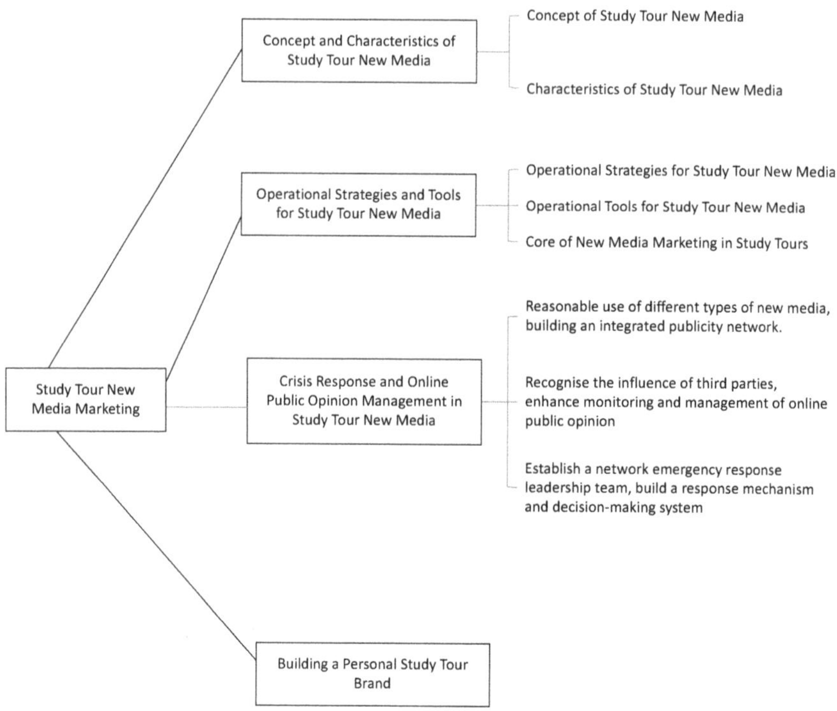

Learning Focus

1. Understand the operational strategies of educational tourism new media.
2. Understand how to respond to and manage online public opinion related to educational tourism new media.
3. Learn how to effectively build a personal educational tourism brand.

Contents

1. Case Introduction

According to relevant news reports, 17 Asian elephants that originally lived in the Xishuangbanna National Nature Reserve in Yunnan set out from the border reserve on 16 April 2021, travelling more than 500 kilometres. By early June 2021, they had approached Kunming, Yunnan. Questions such as the elephants' migration route, the damage caused along the way, and where the elephants were ultimately headed captivated the attention of the nation and turned this event into a small international media hotspot.

International media extensively covered the elephants' migration and, in doing so, inadvertently provided an objective promotion of the geographic and cultural landscape of southern China. This event has been recognised as a new case of China's efforts in biodiversity conservation.

Some domestic social media platforms broadcasted the elephant migration using formats such as livestreams and short videos, without adding any commentary. By fully leveraging the "rules of the game" on overseas social media—characterised by low political content and strong entertainment value—these platforms achieved excellent results.

As such, in the process of telling China's story, Chinese media should skilfully utilise new media forms such as short videos and livestreams, employ communication discourses with transnational cultural commonality (including text, images, animations, etc.), and effectively integrate big data, location-based services (LBS), 5G and other technologies to share China's stories and culture in a soft and appealing manner.

1.1. *Analytical questions*

1. What new media formats did domestic media use in the dissemination of the elephant migration event?
2. Do you think advertisements pushed through new media are more accurate than those released through traditional media (such as newspapers, radio, television, magazines, etc.)? Please explain using the case as a reference.

2. The Concept and Characteristics of New Media in Educational Tourisms

2.1. *The concept of new media in educational tourisms*

Today, social media has become a natural tool for customer management and a powerful means of attracting traffic. Especially in this era of traffic economy, educational tourism enterprises face a key question: should they rely on professional new media teams to manage their product marketing, or should they build and incubate their in-house teams? Regardless of the approach, the ultimate goal remains the same—to leverage new media not just for business support, but for direct business conversion.

UNESCO defines new media as: "media that disseminates information through digital information technology and network-based platforms". In today's world, empowered by advances in digital and network technologies, virtually anything can become new media. For instance, traditionally considered a legacy medium, television can now be classified as new media after digital transformation into digital TV. Likewise, traditional newspapers can be upgraded into digital publications via digital technology and thus become part of the new media landscape. Strictly speaking, any medium based on digital information technology can be regarded as new media. This also means traditional media can evolve into new media through digital upgrading.

The purpose of establishing new media platforms for educational tourisms is to deepen public understanding of educational tourism products. This is achieved by enhancing the credibility of information shared via these platforms, and by integrating multiple media formats to strengthen the role of new media in building and marketing brand identity. This helps educational tourism enterprises and destinations quickly shape their brand image. Through the dissemination of engaging and interactive content, these platforms aim to resonate with audiences and highlight both the tourism and educational value of their products, rather than simply catering to the need for entertainment.

2.2. *Characteristics of new media in educational tourisms*

Compared to traditional media, new media possesses a distinctive style in both its methods of dissemination and content. For educational tourism enterprises seeking to integrate traditional media with innovative marketing approaches, it is crucial to gain a thorough understanding of the characteristics of new media in the context of educational tourisms. These characteristics include:

(1) **Diverse Modes of Communication**
New media offers rapid dissemination, wide reach, and rich content, enabling the release and spread of information with minimal limitations in terms of time, location, ethnicity, identity, or financial status. It boasts strong real-time interactivity, potential for social mobilisation, and significant influence on public opinion—especially evident in the spread of breaking news or key events.

(2) **Surge in Audience Numbers**
New media platforms for educational tourisms now reach audiences across all age groups. According to statistics from iiMedia Research, as of 2019, 43% of China's new media users were aged 31–40, while 29.5% were aged 26–30, reflecting a predominantly younger demographic.

(3) **Massive Volume of Information**
In a new media environment, everyone becomes a content creator—even the smallest individual can build their own educational tourism brand. A large user base combined with fragmented mobile content means that any single message can either be quickly drowned out or rapidly go viral. Some educational tourism providers have used creative new media strategies to encourage participants to share their experiences, thereby enhancing the credibility of marketing messages.

(4) **Coexistence of Entertainment and Serious Content**
Consumers today care more about themselves and whether their personal needs are met. They are more inclined to share unique and entertaining experiences. Therefore, educational tourism new media content must strike a balance between professional, authoritative material and engaging, interactive content in order to resonate with target users.

3. Educational Tourism New Media Operation Strategies and Tools

3.1. *Educational tourism new media operation strategies*

The operation of educational tourism new media is not simply about gaining traffic—the key is to clarify the purpose of the operation. Is it primarily to increase followers? Or to deliver official announcements and promote the company's values? Or is it aimed at activating users and generating leads to drive sales? At different stages, the operation of educational tourism new media will also require content and goal adjustments. Therefore, the operation must follow a comprehensive plan. Through careful content planning and quality traffic exposure, precise sales conversions can be achieved—this is currently the focus of educational tourism companies.

So what exactly does the operation of educational tourism new media involve? How should the work be carried out? Simply put, it mainly consists of five aspects: user research, content creation, channel operation, campaign planning, and community management.

(1) **User Research: Understanding the Needs of the Target Audience**
User research refers to communicating with target users to understand them and build user personas. Only by knowing the users' needs, preferences, and characteristics can we produce and distribute content with precision, pre-validate the effectiveness of marketing campaign plans, improve the return on investment of marketing efforts, and ensure that every cent is spent wisely. Common user research methods include surveys, interviews, and focus group discussions. Before starting the research, it is necessary to clarify its objectives and what information needs to be collected. Regardless of the method used, it is important to target the right users and engage with them, then organise and analyse the data to develop more accurate user personas. Extracting key information from the research data provides a reference point for creating and delivering targeted content and campaign planning. The goal is to address user pain points and focus on what users truly care about. Moreover, the research should identify Key Opinion Leaders (KOLs) whose endorsements can help with promotion. User research also aids in refining the definition of target users to better meet operational needs.

(2) **Content Creation: Producing High-Quality Content That Resonates with the Target Audience**
Content creation includes writing articles and making videos. The key steps are topic selection, headline writing, and leveraging trending topics. This requires knowing what type of content your target audience enjoys, how to write compelling titles, what hot topics are currently trending, and how to connect your content to these trends. It's important to note that a single new media team often faces limitations in resources, influence, and outreach. Therefore, it's essential to go beyond in-house content production by collaborating with other teams and engaging in joint content creation from multiple perspectives. For example, a scenic area may set up a new media editorial department, connect with the industry, segment audiences by experience preferences, and create content through interactive product experiences. In addition, gathering a wide range of relevant

information is a necessary preparation before content creation. Just as scientific research builds on previous work, producing hit articles or viral videos also requires learning from outstanding cases.

(3) **Channel Operation: Driving Traffic to Increase Content Exposure**
Channel operation is responsible for boosting content exposure. In educational tourism new media operations, it is not enough to simply produce quality content—the content must also reach as many target users as possible. No matter how well an article is written or a video is produced, without views, there can be no conversion. For educational tourism enterprises, content that fails to generate traffic lacks value, as the ultimate goal of operations is conversion, and without traffic, conversion is impossible.

In theory, the responsibilities of channel operation mainly include developing an operational plan, analysing traffic sources, and promoting content through various means. Specifically, it can be broken down into three parts: First, plan how many articles, promotional videos, or posters are to be distributed on which platforms over a week or month. Second, analyse which sections of which platforms generate the most effective traffic—such as comparing the traffic yield from Zhihu, Douban, or Weibo. Third, determine how to maximise exposure of published content—such as increasing the number of likes, comments, and views. For example, on We Chat Official Accounts, the most straightforward way to boost visibility is by sharing posts to WeChat groups. Additionally, platforms such as Weibo, Douban, Toutiao, Zhihu, Jianshu, Xiaohongshu, and Yidian Zixun can be used to promote content widely.

At present, educational tourism new media operations still face numerous challenges, with a lack of professionalism and execution capacity among operators being major issues. Fortunately, most major new media platforms have already been established. For example, promoting educational tourism products on Xiaohongshu or Toutiao tends to yield relatively high exposure.

(4) **Campaign Planning: Achieving Marketing Goals Through Organised Events**
Campaign planning typically includes developing an event plan, proposing ideas, executing the plan, and monitoring the process. The workflow involves defining the campaign's objective and target

audience, setting the budget, confirming the event theme, designing specific rules and formats, proposing creative ideas, assigning personnel, and executing each stage according to the plan—while monitoring data and making strategy adjustments as necessary to achieve the set marketing goals such as traffic generation or product promotion. A post-campaign review and summary is also essential. Effective campaigns can act like a "timely rain", bringing significant traffic in a short period with immediate results.

Currently, campaign planning needs to be both innovative and cost-effective, with a strong focus on achieving the campaign's marketing objectives (such as increasing followers, boosting exposure, or generating sales). Educational tourism new media strategies are closely tied to actual educational tourism activities, making it a particularly effective way to increase product visibility. Therefore, new media editors for educational tourisms should work closely with campaign organisers. The campaign theme should align with the product content, include endorsements or insights from Key Opinion Leaders (KOLs), and inspire the public to resonate with and voluntarily share the content. For example, one scenic spot initially struggled with its WeChat Official Account, gaining only about 50 new followers per week and even experiencing occasional losses in followers. Later, by combining "soft promotion online" and "hard promotion offline" to host several events, the account gained over 1,000 new followers in a short time, far more effective than writing 100 articles.

(5) **Community Operation: Enhancing Conversion Rates through Community Engagement**

The core responsibilities of community operation include managing user groups, organising community activities, and converting users through these activities. This involves creating a community, such as a QQ group, WeChat group, or forum, and inviting target users to join. Community rules must be established to maintain order. When members have questions, timely responses should be provided, and their needs and feedback should be collected. Any member who violates community rules should be warned, and repeated violations may result in removal from the group. Periodic activities, such as initiating discussions, should be conducted to maintain engagement and invigorate the community. Community activities can encourage target users to like and purchase educational tourism products or services. It is also essential to maintain relationships with community members on an ongoing basis.

In recent years, many educational tourism companies have recognised the significance of community operation. Many job postings for educational tourism in new media now specifically list community operation as a required skill. This is due to the effectiveness of communities in fostering word-of-mouth brand promotion. Furthermore, many educational tourism enterprises have realised that retaining existing customers is more cost-effective and generates greater value than acquiring new ones. Community operation is often closely linked to campaign operation, with overlapping responsibilities and methods in both roles.

3.2. *Educational tourism new media operation tools*

For new media operators, study does not need to be professional designers or programmers. Still, they should know how to quickly find suitable new media operation tools and how to use them to enhance work efficiency. Therefore, the author has compiled the following new media operation tools for readers' reference.

1. Editing and Layout Tools
   a. 135 编辑器 (https://www.135editor.com/)。
   b. 365 编辑器 (https://www.365editor.com/)。
   c. 6 编辑器 (http://bj.96weixin.com/)。
   d. 速排小蚂蚁编辑器 (http://www.xmyeditor.com/)。
   e. 秀米 (https://xiumi.us/#/)。
   f. 美多编辑器 (http://edit.meiduow.com/)。
   g. 新榜编辑器 (https://edit.newrank.cn/)。
   h. 易点编辑器 (http://www.wxeditor.com/)。
   i. i 排版 (http://ipaiban.com/bianji)。
   j. 91 微信编辑器 (http://bj.91join.com/)。
   k. 排版侠 (http://www.paibanxia.com/)。
   l. 微小宝 (https://editor.wxb.com/)。
   m. 易企微 (http://www.e7wei.cn/bjq.html)。
   n. 微助点微信编辑器 (http://www.videaba.com/)。
   o. 主编编辑器 (http://www.zhubian.com/)。

2. H5 Creation Tools
   a. 易企秀（https://store.eqxiu.com/)。
   b. 兔展（http://www.rabbitpre.com/)。
   c. 码卡（http://maka.im/)。
   d. 人人秀（https://www.rrxiu.net/)。
   e. 初页（http://cloud7.com.cn/)。
   f. 意派（http://www.epub360.com/)。
   g. LiveApp（https://liveapp.site/)。
   h. 秀米（https://xiumi.us/#/)。
   i. 即速应用（http://www.zhichiwangluo.com/)。

3. QR Code Generation Tools
   a. 草料二维码（http://cli.im/)。
   b. 联图网（http://www.liantu.com/)。
   c. 二维码工坊（https://www.2weima.com/)。
   d. 微微二维码（http://www.wwei.cn)。
   e. 第九工场（https://www.909th.com/)。
   f. 互联网二维码（https://www.hlcode.cn/)。
   g. 模板码（http://www.mobanma.com/)。

4. Stock Material Websites
   a. 素材 CNN（http://www.sccnn.com/)。
   b. 大作（http://www.bigbigwork.com/)。
   c. 我图网（http://www.ooopic.com/)。
   d. 花瓣（http://huaban.com/)。
   e. 懒人图库（http://www.lanrentuku.com/)。
   f. 站酷（http://www.zcool.com.cn/)。
   g. 视觉中国（https://www.vcg.com/)。
   h. 千图（http://www.58pic.com/)。
   i. 摄图网（http://699pic.com/)。
   j. 包图（https://ibaotu.com/)。
   k. 昵图网（http://www.nipic.com/)。
   l. 站酷海洛（https://www.hellorf.com/)。
   m. 图虫（https://tuchong.com/)。
   n. picjumbo（https://picjumbo.com/)。
   o. Pexels（https://www.pexels.com/)。
   p. 素材公社（http://www.tooopen.com/)。
   q. pixabay（https://pixabay.com/)。

5. GIF Image Websites
    a. SOOGIF（https://www.soogif.com/）。
    b. 新浪科技 GIF 趣图（http://gif.sina.com.cn/）。
    c. Golden Wol（f https://www.goldenwolf.tv/）。
    d. JULIAN GLANDER（http://glander.co/）。

6. Image Editing Tools
    a. ARKIE（https://www.arkie.cn/）。
    b. 创客贴（https://www.chuangkit.com/）。
    c. 懒设计（https://www.fotor.com.cn/）。
    d. 易图网（https://www.yipic.cn/）。
    e. 图帮主（http://www.tubangzhu.com/）。
    f. 稿定设计（https://www.gaoding.com/）。
    g. 改图宝（http://www.gaitubao.com/）。
    h. 光影魔术手（http://www.neoimaging.cn/）。
    i. UUPOOP（http://www.uupoop.com/ps/）。
    j. 看图网（https://www.kantu.com/）。
    k. GIF5（http://www.gif5.net/）。
    l. 图艺图（http://www.tuyitu.com/）。
    m. 我拉网 GIF 制作（https://gif.55.la/）。
    n. Canva（https://www.canva.cn/）。

7. Survey Tools
    a. 麦客（https://www.mikecrm.com/）。
    b. 金数据（https://jinshuju.net/）。
    c. 问卷星（https://www.wjx.cn/）。
    d. 腾讯问卷（https://wj.qq.com/）。
    e. 我要调查网（http://www.51diaocha.com/）。
    f. CTR 调研社区（http://cn2.ictr.cn/）。
    g. 问智道（http://www.askform.cn/）。
    h. 调查派（https://www.diaochapai.com/）。
    i. 问卷网（http://www.wenjuan.com/）。
    j. 表单大师（https://www.jsform.com/）。

The abovementioned tools cover editing and layout tools, H5 creation tools, QR code generators, and image editing tools. From a practical standpoint, studying new media operators, in addition to mastering basic theory, should also continue learning how to use these tools to steadily improve their application skills.

3.3. *The core of educational tourism new media marketing*

In the rapidly changing field of new media communication, the fundamental challenge of studying new media marketing lies in selecting appropriate content and ensuring it stands out. Therefore, mastering the method of producing creative content becomes particularly important. From a pragmatic perspective, the following four aspects are worth referencing:

(1) **Pinpointing the Right Content Positioning**
Conclusions may be simple, but relationships are complex. The core of content creation in the educational tourism of new media marketing lies in establishing a clear-thinking framework. One should find a unique positioning for the content, develop a distinct style, focus on what the target users care about, and identify its core value. Generally speaking, users choose information based on their perception—meaning, the kind of content a user selects reflects the type of product they are attracted to. Therefore, positioning is key.

(2) **Information Value Must Be Relevant to the Target User**
While new media marketing content may deliver multiple functional values, its fundamental role remains the communication of accurate information. In other words, marketing content must be grounded in objective facts and relevant to consumer needs, behaviours, and contexts—otherwise, it loses its basic communicative value. A educational tourism on new media marketing should focus on content with accurate informational value as a foundation for brand reputation. Product design should be consumer-oriented, emphasise consumer feedback, and consider customer satisfaction the cornerstone of growth. Through new media marketing, companies should showcase service content that highlights personalisation and uniqueness, focusing on content that aligns with consumer needs. Especially during the COVID-19 pandemic, if the information shared is genuinely useful, even a single sentence, image, or clip could go viral. However, fake news is often mixed in, and even if it is quickly debunked, it still disrupts the audience and damages the credibility of the new media source. Thus, useful, accurate, and valuable content must be the basic starting point for all new media content creation.

(3) **Leveraging Trending Topics to Evoke Emotional Resonance**
The previously mentioned elephant migration incident drew widespread attention partly because of the creative and engaging

storytelling techniques used in related articles, such as personification, humour, and a warm tone. On the surface, what spreads appears to be the content itself, but in reality, the emotions conveyed by the content are being shared. The spread of emotion, in turn, accelerates the dissemination of the content and stimulates the audience's impulse to continue sharing it. Therefore, to create emotionally resonant content, one must first be clear about the emotions, sentiments, perspectives, attitudes, and ideas they intend to convey, and identify emotional triggers that can spark empathy and connection with the audience.

(4) **"Viral" Content Creation with Impact**
Neil Postman once said in *Amusing Ourselves to Death* that all public discourse is increasingly presented in the form of entertainment and has become a cultural ethos. In the age of new media marketing, everyone can be a content creator, and content tends to be fragmented, intimate, and entertaining. Fun and entertaining content often spreads "virally," quickly boosting a brand's visibility and awareness. These viral materials are usually presented in humorous or playful formats. However, it is important that such humour remains positive and contributes constructively to the brand's image. In simple terms, "viral" content can be a fun game or a short story that resonates with a broad audience and encourages participation. For example, the "Ice Bucket Challenge" is a classic case of self-propagating content driven by love and awareness. Its rules were designed to be interactive and straightforward: post online, participate in the challenge (pouring ice water), and nominate friends to continue the chain.In the context of educational tourism new media marketing, content creation should focus more on fun, interactive, positive, and educational themes, while avoiding entertaining but poorly thought-out or misleading ideas.

4. Negative Effects of Educational Tourism on New Media and Online Public Opinion Management

Educational tourism new media marketing provides diverse information and spreads rapidly over a wide range. However, this fast-paced communication can also accelerate the spread of vulgar content or online rumours, leading to uncontrolled public opinion. If not managed in time, it will inevitably cause negative emotions to spread and result in serious damage to the brand. Therefore, educational tourism organisations or destinations need to have effective strategies for managing online public opinion crises.

4.1. *Effectively utilise different types of new media to build a multi-dimensional publicity network*

Educational tourism organisations should actively use various forms of new media to engage in timely dialogue and communication with consumers, maintaining their media presence regularly to create a positive environment for the healthy development of educational tourisms. When a public opinion crisis emerges, the organisation should promptly use this multi-dimensional publicity network to guide consumer anxieties, reduce the space for negative information to spread, and prevent the crisis from escalating.

Of course, media platforms are not the only source where public opinion crises may appear. There are also potential crises that have yet to be reported by the media. In such cases, if educational tourism organisations can communicate promptly with stakeholders such as the government, consumers, internal staff, or partners, minor issues can often be resolved before they escalate. Furthermore, organisations must carefully plan the sequence of their communications, taking into account all stakeholders involved in the incident.

4.2. *Value third-party support: Strengthen online public opinion monitoring and management*

Online public opinion monitoring refers to the process of using manual searches or specialised data collection and analysis technologies to monitor, track, and predict targeted information on the internet. This allows the monitoring party to better supervise and respond to relevant content. A standardised workflow is crucial for effective execution. Similarly, educational tourism new media public opinion monitoring follows a structured method, as outlined below:

1. **During the monitoring period**, information from multiple sources is collected and stored in a centralised information database.

2. **The public opinion analysis engine** filters the data in the database. It uses intelligent analytical tools to assess the overall sentiment of the event, such as the proportion of negative, neutral, and positive news reports and the overall sentiment index or "health" of the event.
3. **Timely publication** of relevant data from the public opinion results database.
4. **Staff members** compile a summary of the monitoring outcomes, generate internal briefings, and promptly conduct assessments and submit reports.

4.3. *Establish a public opinion crisis leadership team and response command system*

Responding to online public opinion is a complex and systematic task that requires unified command and coordinated action. Educational tourism enterprises should form a **public opinion crisis leadership team** to serve as an effective response body. Within the authorised scope of the enterprise, they must establish a **standardised and scientific response mechanism and command system**. Without this, there is a high risk of overreaction or insufficient response.

Additionally, **spokespersons** must be trained to improve their communication skills and overall competence, enabling them to guide the public toward rational understanding. A sincere and factual response can significantly help resolve public opinion crises and reduce their negative impact on the enterprise or destination's brand image.

After a public opinion crisis emerges, **active and transparent communication** is crucial to minimising reputational damage. Below are several proven communication principles:

1. **Speak with one unified voice** – Ensure all external communications are consistent to avoid mixed messages from different departments or individuals.
2. **Respond as early as possible** – Address the situation promptly, clarify facts, and guide public discourse positively to prevent escalation due to delayed responses.
3. **Focus on facts, avoid speculation** – Provide timely and factual information, offer a comprehensive overview of the situation, clearly

explain the truth, and publish official statements on the company's website. Respond sincerely to the outcome and decisions made.

5. Building a Personal Educational Tourism Brand

In the era of "everyone is a media outlet" enabled by new media, building a personal brand has become significantly easier. Anyone can register on third-party platforms, share opinions, publish videos, and attract followers. This section focuses on three typical case studies to explore: Branding of educational tourism mentors, branding of educational tourism institutions, and how to build a personal educational tourism brand and use new media to market oneself effectively.

Industry Case Study

"Dr. Yi Talks About Educational tourisms" – A Personal Educational tourism Brand

Background
Since May 2020, due to the impact of the COVID-19 pandemic, Dr. Yi Hong and her team decided to launch a self-media brand under her personal name: **"Dr. Yi Talks About Educational tourisms"**, using platforms like WeChat Channels, WeChat Official Accounts, and Douyin (Chinese TikTok) to establish her personal brand in the educational tourism industry.

1. **Brand Positioning**
   Image: Expert and senior researcher in experiential education.
   Mission: To guide young learners through "knowledge tips" that connect what they learn in books with the real world, helping them see that "*learning happens everywhere in life*".

   Name Interpretation
   The brand name "易博士讲研学" ("Dr. Yi Talks About Educational tourisms") includes three key elements:

   - **Person** – "Dr. Yi" (易博士)
   - **Topic** – "Educational tourisms" (研学)

- **Method** – "Talks" or Sharing (讲) This combination ensures that when the audience connects with the person (Dr. Yi), they also learn about the topic, through the chosen medium. The name is simple, direct, and easy to spread.

2. **Communication Platforms**
WeChat Channel: "易博适"
Douyin Account: "易博士讲产学研" (Dr. Yi Talks Industry-Academia-Tourism Integration)
Baidu Baike Entry: "Yi Hong"
WeChat Official Accounts

- "研学头条" (Educational tourism Headlines)
- "自在研学" (Free Educational tourism)
- "易博士讲研学" (Dr. Yi Talks About Educational tourisms)

Mainstream Media Coverage
Featured in platforms like *Xinhua News Agency*, *Hubei Daily*, *Jingzhou Daily*, etc.

3. **Practical Strategies**

Consistent Quality Content Output
Maintain a steady rhythm of updates (daily, weekly, or biweekly) to ensure regular visibility.

Unique Perspective
Offer original insights by deconstructing everyday knowledge through a educational tourism lens. The content combines **educational value**, **fun**, and **practicality**.

Real-Life Relevance
Discover study-worthy knowledge from daily life. Use lighthearted, accessible content to spark thinking and align with the concept that "*learning happens everywhere in life*".

4. **Theoretical and Content Support**
A knowledge-based content creator cannot consistently produce high-quality material without a solid academic foundation. The "Dr. Yi Talks About Educational tourisms" personal brand is built upon years of research by her educational tourism team.

(1) **Academic Papers:** These include works such as "*Learning Happens Everywhere in Life – Design and Innovation of*

Experiential Educational tourism Teaching" (published in *The Educator*, 2018), and "*Application of the Spark-Based Teaching Method in Experiential Education for Educational tourisms*" (Published in Teacher *Education Forum*, 2019).

(2) (Publications: A range of educational tourism-themed books including the six-volume series *Dr. Bigger's Thinking Cubes* (People's Transportation Publishing House), the five-volume picture book collection *Fun in All Directions* (Wuhan Publishing House), and a parent-child educational tourism title *Why Doesn't Mum Tell Fairy Tales?* (Wuhan Publishing House).

(3) Animated Series: A educational tourism-themed animation titled *Fun in All Directions*.

Li Weihan's Personal Educational Tourism Brand as a Educational tourism Instructor

Li Weihan is a post-90s educational tourism instructor from Luoyang, Henan Province—one of China's 13 ancient capitals. He currently serves at the Yingtianmen Site Museum and Jiuzhouchi Site Park, backed by years of hands-on experience in the field.

Luoyang is exceptionally rich in educational tourism resources related to history, culture, industry, and nature—with historical and cultural resources being its greatest strength. Thirteen dynasties once established their capitals here, leaving behind major ruins from the Xia, Shang, Western Zhou, Eastern Zhou, Eastern Han, Northern Wei, Sui, and Tang dynasties. With over 400,000 surface-level relics and more than 100 museums, Luoyang's heritage provides a powerful educational base. Converting historical resources into engaging curricula is a fundamental skill of a educational tourism instructor.

How can one present ancient history from fresh angles to spark students' curiosity? How can an instructor select relevant historical themes and utilise such resources to build a personal brand? These are the kinds of challenges Li Weihan regularly faces in his professional journey.

1. **"Immersive Experience + Systematic Learning + General Education" as a Course Design Strategy**

   Since 2019, Li Weihan has adopted the design approach of "immersive experience + systematic learning + general education" for educational tourisms. For example, to help students understand Tang Dynasty architecture and history, he studied vast amounts of historical and architectural material and designed various role-play-based educational tourism courses such as *Yingtianmen Construction, Push Open Yingtianmen and Revisit the Tang Empire*, and *Decoding Sui-Tang Brackets, LEGO-Style Exploration of China*. His programmes blend game-based learning with real-world engagement, tapping into students' innate curiosity. Each course follows a challenge-based progression—from encountering a problem to solving it—to support the formation of structured learning. He designs immersive settings and storylines, allowing students to break away from traditional classroom models, explore narratives, solve challenges independently, and ultimately acquire and retain knowledge.

2. **Creating a Tang Dynasty-Themed Series of Educational Tourism Products**
   Over the years, Li Weihan has participated in the development of three major course series. The *Capital Life Guide Series* explores Tang Dynasty lifestyle and aesthetics. The *Great Tang Climbing Ambitions Series* helps learners trace the roots of Chinese civilisation. The *Return of the Tang People Series* centres on role-playing and dramatic reenactments.

3. **Content Ecosystem**
   Li and his team regularly post short videos and conduct livestreams via Douyin and WeChat Official Accounts to share historical and cultural content. Each video is typically under one minute (15, 30, or 60 seconds) and connects historical knowledge with everyday life—such as "How Were Sui and Tang Festivals Different from Today's?" or "How Did Urban Women Celebrate the New Year in the Tang Dynasty?" His content is not limited to students but reaches a broader audience of history enthusiasts.

4. **Awards and Recognition**
   At the beginning of his journey, Li set a personal goal to "avoid explaining history in a conventional way" and to "design curricula that go beyond the classroom". Through immersive storytelling and gamified learning, he received widespread acclaim and offered students truly memorable learning experiences. His course *Push Open Yingtianmen and Revisit the Tang Empire* won first prize in the

Luoyang City Primary and Secondary School Premium Curriculum Design Competition in 2020. In 2021, he was awarded the title "Luoyang Gold Medal Educational tourism Instructor", with media coverage from *Luoyang Daily*, Zhejiang TV, and Henan TV.

In each educational tourism activity, Li breaks away from the role of mere lecturer, positioning himself instead as a designer and facilitator. He guides students to step into historical scenes, take on character roles, and grow through the process of "problem discovery–problem solving". In summary, Li Weihan's journey as a educational tourism instructor reflects how one can build a personal brand in tandem with the evolution of the educational tourism industry.

Zeng Jin's Personal Educational Tourism Brand as a Parent-Child Educational Tourism Expert

In July 2019, Zeng Jin—a seasoned media professional and former deputy editor-in-chief of *The Bund Pictorial*—published a ten-volume educational tourism journal series titled *Travelling China Through Poetry*. She leveraged various platforms, including WeChat official accounts and mainstream domestic media, to promote the series. The content quickly gained traction, garnering over a million media views. With the support of e-commerce channels like Douyin (TikTok China), the book series sparked a sales boom. Between 2020 and 2021, over 200,000 copies were sold, earning Zeng Jin the title of Dangdang's "Influential Author of the Year" for both years.

In February 2022, Zeng Jin developed a board game derivative of the same name. By the summer of 2022, she had begun writing the *China National Parks Book Series*, which further solidified her standing as a leading expert in the educational tourism field. She later rebranded her WeChat and Video Account under the unified name "Xiao Zeng Miya", integrating her three key identities—author, reading expert, and parent-child educational tourism specialist—to reinforce her personal brand. As of February 2023, her WeChat account had attracted over 100,000 followers.

5.1. *Brand positioning*

Image Positioning: Writer, reading expert, and parent-child travel enthusiast.

Brand Philosophy: A love for travel, a love for reading, and a love for life.

Brand Name Explanation: The name centres on the core figure—Zeng Jin—a person with multiple identities. By switching between her three roles, she draws attention from diverse audiences. The brand primarily focuses on the multifaceted needs of parents in reading, travel, and family education. Through Zeng Jin, audiences can see an idealised version of themselves—a modern woman, a loving mother who cherishes life and her child. This identification with her diverse roles inspires followers to grow and evolve, becoming lifelong learners, devoted readers, and thoughtful travellers.

5.2. *Communication platforms*

1. Personal WeChat public account and video account: *Xiao Zeng Miya.*
2. Branded WeChat public accounts, such as *Qixiang Youth*, *Qixiang Elementary School*, *Whole Book Reading for Primary Schools*, and *Big Hands Hold Little Hands*.
3. Reports by mainstream domestic media, including *Xinhua News Agency* and *Shanghai TV Station*.
4. Offline bookstores, such as *Luoxin Bookroom* and *Beijing Dayue Culture*.
5. Schools including *Shanghai Qingpu Pinghe Bilingual School*, *Shanghai Normal University Affiliated Foreign Language Primary School*, *Fudan University Second Affiliated School*, and other well-known primary and secondary schools.

5.3. *Specific practices*

1. Zeng Jin ensures consistent, high-quality content output before every winter and summer holiday or long weekend to maintain visibility. She leverages organic traffic to promote her book series *Travelling China Through Poetry*, ensuring continuous exposure of her personal brand.
2. She uses her social media accounts for fragmented, short-form communication, shaping her persona as a "parent-child travel expert". Her dual identity as a knowledgeable urban woman and a mother, combined with high-quality, personalised, and visualised educational tourism route designs, strongly resonates with parent audiences and strengthens their recognition of her educational and study travel philosophies.

3. **Communication Approach** – Transforming textbook knowledge into "cultural geography". As a reading expert, Zeng Jin is skilled at humanising the geographic knowledge found in textbooks and literary classics. Using the perspective of a modern educator, she brings historical and cultural knowledge to life. Articles like *Travelling China with Sima Qian* and *Travelling China with Lu You* have garnered widespread interest from students and parents alike.
4. **Content Focus** – Designing educational tourism routes incorporating Problem-Based Learning (PBL). While promoting educational tourism concepts, she strongly emphasises localised PBL frameworks that allow students to "learn through travel". For example, she guides primary students through local experiences and leads secondary students in designing research-based travel projects—progressing from garden sketching, calligraphy, and clay figure making to investigating snow leopard conservation and exploring Three Kingdoms battlefields.
5. **Channels** – Integrating private and public traffic streams. She harnesses the influence of celebrities, travel industry CEOs (e.g. Tongcheng Network Technology CEO Wu Zhixiang), and well-known children's authors (like Canran), as well as Douyin influencers with millions of followers, to promote her educational tourism brand. Leveraging her influence in the K-12 education field, she delivers public lectures and uses her identity as a mother to hold community, library, and bookstore-based sharing sessions. Through channels like *Qixiang Elementary School* and *Xiao Zeng Miya*, she consolidates private and public traffic online and offline into social communities for secondary promotion and integration.

5.4. *Theoretical foundation*

Zeng Jin has extensive experience in travel writing. As a mother, she also has a strong grasp of parental psychology. Through her writing, she blends educational psychology, journalism, literature, philosophy, and geography to produce widely circulated articles and publications.

1. She has written over 50 study travel articles with a total online reach in the millions. Examples include column pieces like *Travelling Chengdu with Du Fu* and *Travelling Hangzhou with Su Shi* (*Outlook Oriental Weekly*, 2019), and in-depth features such as *Chongming*

Island—A Sand Island in Constant Growth (*China National Geography*, 2010) and *Weiyu Houses in Northern Guangdong—The Most Elegant Architecture in Rural China* (*China National Geography*, 2011).
2. Zeng Jin has received media coverage from various major platforms, including the documentary *"My Ma Ya"* on Shanghai Radio and Television's News Channel (*Xinwen Fang*).
3. She has published a series of study travel books titled *Travelling China Through Poetry* (10 volumes) with the Publishing House of Electronics Industry.

Through the three case studies above, we can identify four key aspects necessary for building a personal brand in the field of educational tourisms:

1. **Niche Positioning**
   It is crucial to clearly define one's area of expertise. This can be achieved by narrowing the focus based on personal strengths, industry segments, or thematic topics, thereby creating a distinctive and differentiated brand.

2. **Developing an Operational Plan**
   A clear plan is essential, encompassing well-structured content themes, publishing schedules, and platform selection. A strategic roadmap guarantees content consistency and audience engagement.

3. **Consistent Content Output**
   Maintaining visibility and popularity requires ongoing, well-planned content production. A stable, long-term content plan helps reinforce the personal brand and keeps the audience engaged.

4. **Building a Content Matrix and Expanding Channels**
   Leverage utilise multiple promotional platforms and collaborate with other successful personal brands within the industry to amplify visibility. Multi-platform exposure and cross-promotion are essential for enhancing brand influence.

6. Chapter Summary

Overall, under the favourable momentum of cultural and tourism integration, and with the implementation of the national "Double Reduction" policy, the educational tourism industry holds promising prospects. In the era of "Internet+", leveraging new media to promote educational tourism brands and sell educational tourism products has become an essential capability for educational tourism enterprises.

This chapter focused on practical marketing skills, introducing strategies and workflows for operating new media in the educational tourism context, creative approaches to content generation, management of online public opinion, and methods for building a personal educational tourism brand and marketing oneself through new media. In summary, it presents a comprehensive and systematic guide to hands-on new media marketing practices, aiming to support the rapid development of the educational tourism market.

7. Post-chapter Exercise

1. Based on real-life examples, discuss how empathetic content should be communicated in educational tourism curriculum design.
2. Based on real-life examples, briefly explain how to strike the right balance in terms of scale, tone, and boundaries when creating "viral" breakout content.
3. Do you have any plans to build a personal brand? Try to share your thoughts and ideas.

Index